MATHEMATICS EDUCATION IN KOREA

Volume 2: Contemporary Trends in Researches in Korea

SERIES ON MATHEMATICS EDUCATION

Series Editors: Mogens Niss *(Roskilde University, Denmark)*
Lee Peng Yee *(Nanyang Technological University, Singapore)*
Jeremy Kilpatrick *(University of Georgia, USA)*

Series on Mathematics Education Vol. 11

MATHEMATICS EDUCATION IN KOREA

Volume 2: Contemporary Trends in Researches in Korea

Edited by

Jinho Kim
Daegu National University of Education, Korea

Inki Han
Gyeongsang National University, Korea

Mangoo Park
Seoul National University of Education, Korea

Joongkwoen Lee
Dongguk University, Korea

World Scientific

NEW JERSEY · LONDON · SINGAPORE · BEIJING · SHANGHAI · HONG KONG · TAIPEI · CHENNAI

Published by

World Scientific Publishing Co. Pte. Ltd.

5 Toh Tuck Link, Singapore 596224

USA office: 27 Warren Street, Suite 401-402, Hackensack, NJ 07601

UK office: 57 Shelton Street, Covent Garden, London WC2H 9HE

Library of Congress Cataloging-in-Publication Data
Mathematics education in Korea / edited by Jinho Kim, Daegu National University of Education, Korea, Inki Han, Kyeonsang National University, Korea, Mangoo Park, Seoul National University of Education, Korea, Joongkwoen Lee, International Relation Subcommittee in Korea Society of Mathematical Education, Korea.
    p. cm -- Series on mathematics education ; vol. 7, vol. 11
  Includes bibliographical references.
  ISBN 9789814405850 (v. 1)
  ISBN 9789814525718 (v. 2)
  1. Mathematics -- Study and teaching -- Korea (South). I. Kim, Jinho, editor. II. Han, Inki, editor. III. Park, Mangoo, editor. IV. Lee, Joongkwoen, editor.
  QA11.2 .M27785 2013
  510.71/05195
                      2012518999

British Library Cataloguing-in-Publication Data
A catalogue record for this book is available from the British Library.

Printed in Singapore

OVERVIEW

The Contemporary Research Trends and Practices in Korea is the second volume following the *Mathematics Education in Korea: Curricular and Teaching and Learning Practices* that was published in 2013. These two books aim to address the theory and practice of mathematics education and research trends in Korea.

To those outside of Korea, Korean mathematics education and research were rather unknown. Theories and practices of Korean mathematics education and research that are presented in the two books will contribute to the 'globalization' of Korean mathematics education. Many prominent mathematical achievements of Korean students in TIMSS, PISA, and International Mathematics Olympiads have drawn the attention of mathematics educators and researchers outside of Korea. Although the history of Westernized mathematics education in Korea is not so long, research on mathematics education is vigorous due to national interest and enthusiasm toward mathematics and mathematics education. By successfully hosting the ICME-12, many mathematics educators and researchers pay attention to Korean mathematics education.

The first professional academic institution, the *Korean Society of Mathematical Education,* was founded in 1962 and the society published first professional journal *The Mathematics Education* in 1963. *The Mathematics Education* has been issued successfully and has published 145 issues up to May, 2014. Now, there are many academic and professional organizations that publish various journals on mathematics education.

Mathematics education research has been done by members in relevant professional academic societies, and the results have been presented mainly to professional journals. Research areas in mathematics education vary, ranging from the level of kindergarten to that of university. Research also includes philosophical and psychological aspects, the theory and practice of teaching and learning mathematics, and pre-service and in-service mathematics teacher education and so forth.

The characteristics of the trends of mathematics education in Korea are represented by their 'diversity.' Even though it is impossible to systematically analyze the diversity as a whole, to extract such diversity and describe some of the cross-sections of the mathematics education in South Korea would be a meaningful attempt to understand mathematics education in Korea.

This book consists of four parts. *Part 1: Mathematics Teacher Education* deals with pre-service and in-service mathematics teacher education from primary to secondary levels. *Part 2: Special Mathematics Education* includes mathematically gifted education in Korea and using technology in secondary mathematics education. *Part 3: Development of Mathematics Education* consists of the analysis of the development of Korean mathematics education and suggest the direction of development of kindergarten mathematics curriculum. And *Part 4: Glimpse of Korean Mathematics Education* consists of the discussions from the perspectives of foreign mathematics educators.

Part 1: Mathematics Teacher Education composes Chapters 1, 2, 3, and 4.

In Chapter 1: *The Elementary teacher education programs with a mathematics concentration,* Dr. Pang suggests that Korean students' high scores of mathematics achievements in the international studies such as TIMSS and PISA attribute to Korean teachers' competency. This chapter introduces elementary teacher education programs with a mathematics concentration in Korea and describes the process of teacher recruitment along with the sample items used in the recent National Teacher Employment Tests. Finally, the author suggests challenges that elementary teacher education programs have faced with in Korea.

In Chapter 2: *The pipeline to becoming an elementary teacher and opportunities to learn mathematics for teaching in Korea,* Dr. Kim introduces the pipeline to becoming an elementary teacher in Korea, from entering the teacher education program to getting advanced teacher certification. The author also explains curriculum of teacher education programs and professional development and finally closes with some issues to offer enough opportunities for all teacher candidates to develop their mathematical knowledge and skills for teaching mathematics.

In Chapter 3: *Student teaching in mathematics education at the secondary level in Korea*, Dr. Lee and Ms. Lee provide the student teaching system in mathematics education at the secondary level in Korea. They introduce the background information of the student teaching system in Korea. They also articulate the definition and goals of student teaching and explain the participants of the student teaching process and their roles. As an example, they then provide the timeline, main activities, and products throughout student teaching required at Ewha Womans University, a private female university located in the capital city, Seoul. Finally, they give a discussion on how the student teachers are evaluated and present research on the perception of student teaching.

In Chapter 4: *Master teacher's role in professional development and the mentor teacher*, Dr. Kim introduces the newly established master teacher system in Korea. The author discusses the purpose of the master teacher system, the role of the master teacher, and some experiences of a master teacher specializing in mathematics curriculum and instruction.

Part 2. Special Mathematics Education composed of Chapters 5, 6, 7, and 8.

In Chapter 5: *Gifted education in Korea*, Dr. Park introduces the gifted education in South Korea. He argues that one of the main characteristics of gifted education in Korea is that the Korean government has a systematic and centralized system. He claims that there are still many challenges to increase the size and scope and enhance the quality of programs in gifted education. As an overview of gifted education, he reviews the goal and policy, history and present status, identification and selection process, institutions and programs, teacher education, and support systems of gifted education in Korea.

In Chapter 6: *Qualifications of teachers in specialized secondary STEM schools in Korea*, Dr. Choi and Ms. Whang provide information about teachers in secondary specialized STEM schools such as Science High Schools (SHSs) and Science Academies (SAs) in Korea. They describe that SAs and SHSs offer various professional development programs in order to improve quality of teaching for gifted students in STEM. They suggest that school administrators should pay a special attention to retain high quality teachers as well as clarify expectations for their teacher qualifications.

In Chapter 7: *Mathematics camp for mathematical Olympiad*, Dr. Lee describes that South Korea focuses education for the gifted students by holding educational camp programs during the summer and winter vacations provided by mathematics and science Olympiads and gifted development centers. He explains that the educational camp programs are used for selecting new students for science high schools. He also describes the KMO (Korean Mathematical Olympiad) camp program. He discusses the structure of the mathematics camp and the emotional experience encountered by the participants of the mathematics camp.

In Chapter 8: *Technology use in secondary mathematics education in Korea*, Dr. Son introduces a survey on the use of technology in the secondary mathematics education in Korea. He analyzes the transition of stance of technology use in the secondary mathematics curricula and in the secondary mathematics textbooks. Finally, he presents the survey results on the characteristics of the in-service teacher trainings in relation to the use of technology.

Part 3: Development of Mathematics Education consists of Chapter 9 and 10.

In Chapter 9: *Development of mathematics education in Korea: The role of the Korean Society of Mathematical Education*, Dr. Han, Dr. Ryang, and Dr. Kim argue that the role of the *Korean Society of Mathematical Education* has been central in the development of mathematics education in Korea. They review how and what the society has worked for the development of mathematics education of Korea. They provide a reflective and evaluative review and suggest a future direction of mathematics education of Korea.

In Chapter 10: *Curriculum reform and research trends in early childhood mathematics education in Korea*, Dr. Lee introduces contemporary curriculum reform and research trends in early childhood mathematics education in Korea. She examines the Nuri curriculum, a recently developed and implemented national-level early childhood curriculum and research trends in early childhood mathematics education in the recent four years (2010–2013). Finally, she provides some insights into the agenda for future research to benefit from recent curriculum changes and offers suggestions for improving the quality of the early childhood mathematics education curriculum.

Part 4: Glimpse of Korean Mathematics Education consists of Chapter 11 and 12.

In Chapter 11: *Educating for the future: An outsider's view of South Korea mathematics education*, Dr. Albert provides observations about challenges and opportunities regarding the strengths of South Korea mathematics education program. She majorly argues that current reform-based instruction is to educate for the future and not the present. She suggests three qualities of the current reform movement support this argument: retooling teaching and learning to develop creativity, problem solving and communication.

In Chapter 12: *The Common Core Mathematics Standards and implications for the South Korean curriculum*, Dr. Colen, Ms. Colen, and Dr. Kim examine the meticulous process in crafting the Common Core Mathematics Standards and the Standards for Mathematical Content and Practice. They explore some plausible implications for the South Korean mathematics curriculum.

In this book, the authors suggest analytic discussions in a detailed level as well as general overviews of mathematics education and mathematics education research in Korea. Although this book shows only some aspects of Korean mathematics education, the authors wish to address some core aspects of Korean mathematics education to those who want to know Korean mathematics education and research, and have an opportunity to share the insight on mathematics education research.

Jinho Kim, Inki Han, Mangoo Park, and Joongkwoen Lee

CONTENTS

Part 4 Glimpse of Korean Mathematics Education

CHAPTER 1

ELEMENTARY TEACHER EDUCATION PROGRAMS WITH A MATHEMATICS CONCENTRATION

JeongSuk Pang

Department of Elementary Education (Mathematics Education),
Korea National University of Education
Gangnae-myeon, Cheongwon-gun, Chungbuk 363-791, South Korea
E-mail: jeongsuk@knue.ac.kr

After the release of international comparative studies on student achievement, there has been increased interest in the competencies of teachers across countries. Given that teachers' competencies are mainly developed through teacher education programs, many researchers have paid attention to how pre-service teachers are professionally prepared in these programs and what kinds of knowledge, skills, and competencies they are expected to equip at the end of teacher education program. In responding to this call, this chapter introduces elementary teacher education programs with a mathematics concentration in Korea. The chapter first begins with the general description of elementary teacher education programs, followed by an overview of those with a mathematics concentration. It then describes the process of teacher recruitment along with the sample items used in the recent National Teacher Employment Test. This chapter closes with a discussion of the challenges facing elementary teacher education programs.

1 Introduction

To teach elementary students in Korea, one must be certified as an elementary school teacher which requires the completion of specific coursework set by one of 13 universities offering elementary teacher education programs. Among them, 10 national universities, which are geographically spread out across the country (one or two per province), are specifically designed to prepare elementary teachers (Grades 1-6); one national university offers education for teachers for all grade-levels (Grades K-12). The other two comprehensive universities serve for general colleges students enrolled in programs across various disciplines

1

(e.g., humanities, social sciences, engineering, and other professional schools), along with an elementary teacher education program in its college of education. Compared to the 10 national universities which serve to prepare only elementary teachers, the other three universities serve a relatively small number of students concentrating in elementary education.

This chapter provides an overview of elementary teacher education programs with a concentration in mathematics. It first describes elementary teacher education programs in general, mainly focusing on common curricula across the universities. It then gives an overview of courses required for pre-service elementary teachers who choose mathematics as their concentration. It then provides some characteristics of teacher recruitment and employment, along with the sample items used in the recent National Teacher Employment Test. This chapter closes with a discussion of the issues and challenges specific to elementary teacher education programs.

2   Elementary Teacher Education Programs in General

2.1 *Overview*

The coursework requirements of elementary teacher education programs consist of general studies, pedagogical preparation, subject matter preparation, and fieldwork experience. First, the purpose of general studies is to broadly educate pre-service teachers and to develop a commonly held foundation of knowledge for them in humanities, foreign languages, social sciences, natural sciences, and arts during their first-year program. Second, pedagogical preparation offers students broad perspectives on theories of learning, curricula, and general methods or strategies that could be used in classrooms (e.g., educational psychology, classroom management, and educational philosophy). Third, subject matter preparation offers learning opportunities for pre-service teachers about how to teach each subject matter effectively. This includes moral education, Korean language education, social studies education, mathematics education, science education, physical education, music education, fine art education, practical arts education, English education, computer education, and integrated subject education. Lastly, the fieldwork experience offers teaching opportunities for pre-service teachers in elementary classrooms.

Upon the acceptance of admission to the university, pre-service teachers choose one subject matter for their concentration. Beyond the common courses detailed above, pre-service elementary teachers are required to take their concentration courses during their junior and senior years. Given that the required credit hours in concentration courses range between 20 and 22 out of the total credit hours requirement for the completion of teacher education program of between 140 and 147, about 85% of the coursework is quite the same regardless of the concentration within the university.

2.2 *Mathematics Courses in the General Studies*

During the four-year elementary teacher education program, regardless of the concentration, all pre-service elementary teachers are expected to take mathematics courses (2-3 credit hours) as a part of their general studies requirement. The scope of mathematics covered by these courses varies, but all of them provide a mathematical foundation for pre-service elementary teachers. Table 1-1 summarizes mathematics courses, as a general studies requirement, offered by six elementary teacher education programs.

As shown in Table 1-1, three universities (i.e., B, C, and E) require pre-service teachers to take one mathematics course as a core course for their general studies requirement, whereas the other three universities (i.e., A, D, and F) do not set up a mathematics course as a core course. Upon close examination of elective courses (see * in Table 1-1); however, without exception, even these universities require pre-service teachers to take at least one mathematics course. For instance, pre-service teachers in the institution A are required to take the course *World of Mathematics* by the regulation, and pre-service teachers in the institution D must take either the course *Mathematics in Everyday Life* or *Cultural History of Mathematics*. Moreover, most institutions offer additional courses (see *italics* in Table 1-1) as electives. For instance, pre-service teachers in the institution B must choose one course from either of the two mathematics or four science courses. Similarly, pre-service teachers at institution D must choose one course from either of three mathematics or three science courses.

Table 1-1. Mathematics Courses in the General Studies

Institution	Core Courses (Credit Hours)	Elective Courses (Credit Hours)	
A		* World of Mathematics (3) *Mathematics and Information Society* (3) *Mathematics and Culture* (3)	
B	Foundation of Mathematics (3)	*World of Mathematics* (2) *History of Mathematics* (2) Four other courses in natural sciences	select one
C	Foundation of Modern Mathematics (2)	*Nature of Mathematics* (2) Four other courses in natural sciences	select one
D		* Mathematics in Everyday Life (2) * Cultural History of Mathematics (2)	select one
		Understanding of Modern Mathematics (2) *Recreational Mathematics* (2) *Technology and Mathematics* (2) Three other courses in natural sciences	select one
E	Understanding of Mathematics (2)	*Mathematics in Everyday Life* (3) Five other courses in natural sciences	select one
F		* Understanding of Mathematics (2) * History of Mathematics by Story (2)	select one

Some institutions do not limit the choice from either mathematics or science course but provide a variety of options from which pre-service teachers can choose. For instance, institution C offers 26 extra courses, two of them specifically mathematics courses: *Everyday Life and Mathematics* and *Paper-folding Mathematics*. Similarly, institution D also offers three extra mathematics courses, *Teaching Mathematics for Low Achievers*, *Mathematical Manipulative Materials and Playing with Mathematics*, and *Teaching Mathematics for Gifted Children*.

To summarize, the nature of mathematics courses, as a part of the general studies requirement, varies by institution, but such courses play a significant role in providing an initial opportunity for pre-service elementary teachers to acquire a mathematics foundation[4]. In other words, while taking mathematics courses as either a core or an elective course, pre-service teachers are expected to develop basic mathematical literacy.

2.3 *Mathematics Education Courses in the Subject Matter Requirement*

In the second-year of their education, all pre-service elementary teachers, regardless of their concentration, are required to take two core courses about teaching elementary mathematics, for a total of five credit hours. Table 1-2 summarizes mathematics education courses offered by the six elementary teacher education programs.

Even though the titles of the courses vary across institutions, the nature of these courses is quite similar. The first course of two credit hours introduces theories of teaching elementary mathematics, including the purpose, history, philosophy, and psychology of mathematics education, and the instructional principles of teaching elementary mathematics, as well as problem-solving and reasoning[5,17]. In teaching these courses, instructors employ various instructional formats such as pre-service teachers' presentation, workshops, or cooperative activities, but most instructors use a lecture-oriented style to cover the many theories and topics within a limited time[12,25].

The second course of three credit hours covers practical issues related to teaching elementary mathematics[2,23]. For instance, this course deals with teaching methods and materials tailored to the specific content-domain of mathematics (e.g., number and operations, measurement, and geometry). Most institutions assign a single three-credit-hour course but the institution F, as seen in Table 1-2, divides it into three specific content-domain courses with a total of five credit hours. Because of the nature of this course, a variety of instructional formats are employed. A common approach is to make a connection between the university coursework and elementary mathematics instruction. For instance, an instructor may ask pre-service teachers to analyze elementary mathematics textbooks, to plan a lesson, and to teach mathematics in an elementary school classroom during their student teaching[19]. By discussing a video-taped lesson, pre-service teachers have an opportunity to explore in what ways the various instructional theories they learned in their first course could be applicable in teaching mathematics in their classrooms[15,20].

Because pre-service elementary teachers are required to take at least two courses for each subject matter (both theory and practice) as well as several courses to develop their skills in music, fine arts, and physical education, the fulfillment of the subject matter requirement is quite demanding for them.

Table 1-2. Mathematics Education Courses in the Subject-Matter Requirement

Institution	Core Courses (Credit Hours)
A	Theory for Teaching Elementary School Mathematics (2)
	Practice for Teaching Elementary School Mathematics (3)
B	Elementary Mathematics Education (2)
	Study of Teaching Materials and Methods for Elementary Mathematics (3)
C	Understanding of Elementary Mathematics Education (2)
	Practice of Elementary Mathematics Education (3)
D	Mathematics Education I (2)
	Mathematics Education II (3)
E	Elementary Mathematics Education (2)
	Study of Teaching Materials and Methods for Elementary Mathematics (3)
F	Introduction to Elementary Mathematics Education (2)
	Study of Teaching Materials in Number and Operations (2)
	Study of Teaching Materials in Relation and Measurement (2)
	Study of Teaching Materials in Geometry (1)

To summarize, the nature of mathematics education courses, as a part of subject matter requirement, is similar across institutions. Taking two courses in teaching elementary mathematics, one about theory and the other about practice, enables pre-service teachers to understand the key concepts behind elementary mathematics education and to learn how to teach mathematics to elementary students. These two courses play a critical role for pre-service teachers whose concentration is not mathematics, as these courses are the last chance to develop a basic understanding of how to teach mathematics to elementary students.

2.4 *The Fieldwork Experience*

In the four-year teacher education programs, pre-service elementary teachers spend eight to ten weeks for a total of four to five credit hours in elementary school classrooms as a part of their fieldwork. Table 1-3 summarizes the types and characteristics of fieldwork designed by the six elementary teacher education programs.

The types of fieldwork seem slightly different from one institution to another, but all institutions offer three types of fieldwork. First, the *observation practicum* is usually offered in the first or second year of the teacher education program, even before offering subject matter requirement courses. This practicum is designed to provide an apprenticeship opportunity by observing in-service teachers' teaching for one or two weeks. Thus, it is usually evaluated as Pass or Fail (e.g., institutions B, C, E).

Second, the *teaching practicum* is usually offered in the third year of the teacher education program. This practicum is intended to provide teaching experience of multiple subjects in elementary school classrooms. Because of its importance, the teaching practicum takes at least half of the entire fieldwork period. It is also usually divided into *Teaching I* and *Teaching II* to give pre-service teachers an opportunity to teach multiple subject matters at different grade levels in an elementary school.

Third, the *administrative practicum* is usually offered in the final year of the teacher education program. This practicum is intended for them to deal with practical affairs. In this practicum, pre-service teachers are expected not only to teach subject matters to students but also to learn how to handle official documents. In addition to these common types of experiences, some institutions (e.g., institution E) offer unique student teaching experience in rural areas (e.g., farming or fishing communities, remote places).

Table 1-3. Types and Characteristics of Student Teaching Experience

Institution	Type	Credit Hour	Week	Period
A	Observation & Teaching	2	4	3-2*
	Teaching & Working Practice	2	4	4-1
B	Practicum I: Observation	Pass/Fail	1	2-1
	Practicum II: Teaching	1	4	3-2
	Practicum III: Synthesis	2	4	4-1
	Practicum IV: Cooperation**	1	.	4-2
C	Observation	Pass/Fail	1	2-1
	Participation	1	2	2-2
	Teaching I	1	2	3-1
	Teaching II	1	2	3-2
	Working Practice	1	2	4-1
D	Observation I	1	1	1-2
	Observation II	1	2	2-1
	Teaching I	1	3	3-2
	Teaching II	2	4	4-1
E	Observation I	Pass/Fail	1	1-1
	Observation II	Pass/Fail	1	2-2
	Teaching in farming/fishing communities	1	2	3-1
	Teaching I	1	2	3-2
	Teaching II & Working Practice	2	4	4-1
F	Observation	1	2	2-2
	Teaching I	1	2	3-1
	Teaching II	1	2	3-2
	Working Practice	1	2	4-1

* 3-2 means the *second* semester of the *third* year in a teacher preparation program
** The practicum IV in the institution B requires pre-service teachers to do educational services in elementary schools for more than 40 hours over the four years. The fulfillment of this requirement is confirmed by the second semester of the fourth year in the teacher education program

During the fieldwork, cooperating teachers usually assess the performance of pre-service elementary teachers. The evaluation includes attendance, general attitude, classroom management, teaching performance, and reflective journal writing. The specific criteria vary by the characteristics of practicums. For instance, the observation practicum assesses mainly the details of observation of each lesson, whereas the teaching practicum evaluates the planning, implementing, and reflection of each lesson.

All institutions require pre-service teachers to participate in community services. A common requirement includes educational services for 30 hours over the four years. For instance, pre-service teachers participate in extra-curricular activities in elementary schools, assist children from low-income families, or provide academic support for low-achievers. This is evaluated as Pass or Fail.

To summarize, the fieldwork experience aims to develop pre-service teachers' practical knowledge and skills by observing in-service teachers' instructions, teaching multiple subject matters in elementary classrooms, and playing a role as elementary school teachers as well as providing service in substantial educational contexts. Practical teaching competence of pre-service teachers is emphasized in teacher education programs[6,22] and the knowledge needed for mathematics teaching is nested within teaching practice itself[1]. In fact, the recent National Teacher Employment Tests emphasize the competent performance of teaching in a micro-teaching setting. For these reasons, the fieldwork experience component has been strengthened throughout teacher education programs.

3   Elementary Teacher Education Programs with a Mathematics Concentration

As mentioned above, about 85% of the elementary teacher education program within an institution is similar for all pre-service teachers. The only difference happens in pre-service teachers' choices of concentration. Every teacher education program has its own requirements for pre-service teachers who choose mathematics as their concentration. During teacher education programs, pre-service teachers are required to take seven to ten courses for a total of 20 to 22 credit hours during their junior or senior years. Table 1-4 summarizes mathematics concentration programs of the six teacher education programs.

Table 1-4. Mathematics Concentration Programs

Institu-tion	Core Courses (Credit Hours)	Elective Courses (Credit Hours)	
A	Elementary Algebra (3) Elementary Geometry (3) Study and Practice of Elementary Mathematics Education (3)	Foundation of Elementary Mathematics (3) Materials of Elementary Mathematics (3)	select 1
		Elementary Analysis (3) Elementary Discrete Mathematics (3) Elementary Mathematics Education and Instructional Media (3)	select 1
		Elementary Statistics (3) Psychology of Learning Mathematics (3)	select 1
		Understanding of Elementary Mathematics Instruction (3) Histories of Mathematics and Mathematics Education (3)	select 1
B	Algebra (3) Geometry (2) Analysis (3) Probability and Statistics (3) Study of Elementary Mathematics Education (2) Psychology of Mathematics Education (2)	History of Elementary Mathematics Education (2) Technology for Elementary Mathematics Education (2)	select 1
		Evaluation of Elementary Mathematics Education (2) Curriculum of Elementary Mathematics Education (2)	select 1
		Learning Guide of Elementary Mathematics (2) Set Theory and Topology (2)	select 1
C	Problem Solving in Elementary Mathematics (2) Didactics of Elementary Mathematics Education (2) Research Methodology of Elementary Mathematics Education (2) Elementary Mathematics Curriculum and Evaluation (2)	Algebra and Elementary Mathematics Education (2) Analysis and Elementary Mathematics Education (2) Statistics and Elementary Mathematics Education (2)	select 1
		Geometry and Elementary Mathematics Education (2) Elementary Mathematics Education Using Multimedia (2) Mathematical Discovery (2)	select 1
		History of Mathematics and Mathematics Education(2) Mathematics Education for Young Children (2) Mathematics Education for Gifted and Retarded Students(2)	select 1
		Study on Elementary Number and Operations Education (2) Study on Elementary Geometry and Measurement Education (2) Study on Elementary Pattern and Statistics Education (2)	select 1
		Recreational Mathematics(2), Mathematics Games and Puzzles(2), and other courses	select 2
D	Practice Using Manipulative Materials in School Mathematics (3) Didactics of Mathematics	Cultural History of Mathematics II (3) Mathematics in Real Life II (3)	select 1
		Introduction to Algebra (3) Probability and Statistics (3)	select 1

	(3) Materials of Mathematics (3)	Teaching Mathematical Problem Solving (3) Psychology of Learning Mathematics (3)	select 1
		Introduction to Geometry (3) Set Theory and Topology (3)	select 1
E	* All courses are compulsory Analysis and Function Education (2) Number Theory and Operation (2) Modern Algebra Education (2) Understanding and Practice of Mathematics Education (3) Statistics and Statistical Education (3) History and Basic Theory of Mathematics (2) Elementary Geometry (3) Educational Theory of Exceptional Student and Evaluation (2) Computer and Mathematics Education (2)		
F	Didactics of Mathematics Education (3) Cultural History of Mathematics (3) Algebra (3) Study of Mathematics Education (3)	Analysis (3) Discrete Mathematics (3)	select 1
		Geometry (3) Set Theory (3)	select 1
		Seminar on Mathematics Education (2) Materials of Elementary Mathematics (2)	select 1

All but the institution E offer both core courses and various elective courses. This is intended for pre-service teachers to deepen their knowledge in mathematics and mathematics education. Note that mathematics concentration programs usually consist of two types of courses: pure mathematics and mathematics education courses. Traditionally, mathematics concentration programs had mainly focused on pure mathematics (e.g., number theory, set theory, algebra, geometry, probability and statistics, analysis, and topology) with a few courses of mathematics education (e.g., didactics of elementary mathematics education and teaching materials in elementary mathematics). In other words, content knowledge held a dominant position over pedagogical content knowledge in elementary teacher education programs. This may have resulted from the cultural expectation that even an elementary school teacher who teaches all subject matters needs to have at least a strong content knowledge in the mathematics concentration.

However, the emphasis on pure mathematics has been gradually changed such that pure mathematics courses are now connected to elementary mathematics education, or at least tailored for elementary school teachers. It is evidenced by the titles of courses such as *Elementary Algebra* in the institution A, *Algebra and Elementary Mathematics Education* in the institution C, or *Modern Algebra Education* in the institution E. This is an attempt to differentiate an elementary teacher education program from either a secondary teacher education program or a program for university students who major in mathematics. In fact, it has been underscored that mathematics content courses should be designed as a way for pre-service teachers to deepen their understanding of mathematics content taught in elementary school, rather than as an advanced level of college mathematics, such as algebra or calculus as a discipline[25]. The most dramatic change occurs in institution C in which none of the courses are offered as pure mathematics. A middle ground example is at institution D in which all three core courses deal with mathematics education. Even the most frequent pure mathematics course of algebra, which is offered as a core course in most universities, is offered as an elective course in institution D.

In general, most institutions maintain a good balance between pure mathematics courses and mathematics education courses. Pure mathematics courses are intended for pre-service elementary teachers to deepen their disciplinary knowledge of mathematics, whereas mathematics education courses are intended to enrich their pedagogical content knowledge. This is in sharp contrast to secondary mathematics teacher programs in which content knowledge is significantly dominant in comparison to pedagogical content knowledge[7,8].

Another noticeable characteristic is that mathematics education courses are more varied than pure mathematics courses. In other words, pure mathematics courses are rather fixed, reflecting the sub-disciplines of mathematics such as algebra, geometry, and analysis, whereas mathematics education courses contain a wide variety of foci from fundamental courses dealing with general theories (e.g., Psychology of Learning Mathematics or Curriculum of Elementary Mathematics Education) to ordinary courses related to teaching elementary mathematics (e.g., Elementary Mathematics Education and Instructional Media) to advanced courses related to research on mathematics instruction (e.g., Study and Practice of Elementary Mathematics Education).

Various mathematics education courses may be offered for three reasons. First, high quality elementary mathematics instruction depends more on pedagogical content knowledge than on disciplinary knowledge in mathematics itself. To be clear, because of high entry selectivity— only top students from high school graduates can get an admission from elementary teacher education programs—student mathematical knowledge is high at the outset of teacher education program[12]. The concern then is how to foster student pedagogical content knowledge during the teacher education programs. Second, elementary teacher education programs have more faculty members with doctoral degrees in mathematics education than do those in pure mathematics. Interestingly, the opposite is true for secondary mathematics teacher education programs. Third, pre-service teachers are regarded as active participants who are able to reflect on their teaching rather than to only deliver what they have learned from the university course into their classrooms[16,19]. In fact, the advanced courses dealing with research on mathematics instruction have been only recently added to the elementary teacher education programs. Pre-service teachers are expected to consider reflective practice as natural in learning how to teach mathematics[8].

4   Teacher Employment

4.1   *Overview*

To become an elementary school teacher, one needs to be certified from one of 13 specific universities offering elementary teacher education programs. Because these universities are regarded as top universities in Korea, the entry selectivity is relatively high. The profession of teaching is one of the most desirable jobs in Korea because of its security, stability, and social respect. Once one becomes a teacher, she or he can teach for a quite long period of time without dismissal. Despite large class sizes, other occupational conditions are favorable for Korean teachers (e.g., teaching hours and annual salary increases) when compared to U.S. teachers[18]. Furthermore, being a teacher is traditionally valued and respected. As a result, only top students from high school graduates can gain admission to elementary teacher education programs. Upon the successful completion of the coursework requirements, the teaching certificate is conferred to these pre-service teachers.

Until early 1990s, a graduate from an elementary teacher education program could get a position as a teacher at one of public elementary schools in the province in which the university is located, without taking any examination. Since then, however, the policy of no required examination has been changed and every graduate has to pass the National Teacher Employment Test (NTET) to be employed as a public elementary school teacher. Correspondingly, there is no restriction for a pre-service teacher to apply for a teaching position in other provinces regardless of the location of his or her university. As a result, the competition is becoming intense, especially to be a teacher in main metropolitan cities such as Seoul or Daejeon. Because the number of elementary school student is decreasing due to the low birth rate in Korea, with the steadfast popularity of the teaching profession, the NTET has become even more competitive.

The NTET for a public elementary school teacher consists of two steps: (a) a written essay related to the teaching profession and items related to the entire elementary school curriculum (both short-answer and constructed response items), and (b) an in-depth interview and performance test dealing with pre-service teachers' skillful teaching of subjects including English. As a result, because of the high entry selectivity to the elementary teacher education program and the NTET, the overall quality of teachers is outstanding.

4.2 *The Written Test of the NTET*

The written test of the NTET consists of writing an essay related to the teaching profession (20 points) and writing responses to 22 items related to all subject matters taught in elementary schools (80 points). In terms of subject matter items, the distribution of points varies. For instance, 11 out of 80 points are allotted to the test items on mathematics, with the highest point distribution given to items on the Korean language.

The test items are usually embedded in various contexts such as planning or implementing a lesson, and assessing pre-service teachers' subject matter knowledge and pedagogical content knowledge. For instance, the following Figure 1-1 is one sample item used in the 2014 NTET.

The following is a part of lesson dealing with problem-solving.

> Teacher: Solve the problem.
>
> *<Problem>*
>
> *Soccer games are planned among 7 classes in Seul-gi School. Given each class is to play with the other classes once, figure out the number of total games.*
>
> (Students solve the given problem)
>
> Teacher: Let's talk about how you solved the problem.
>
> Jae-hi: I drew a picture. After connecting 7 classes and counting the connected lines, I found out 21 games.
>
> Teacher: Good! What do you think of Jae-hi's method?
>
> Min-su: At first I tried to solve like that, but I realized I might make a mistake in counting all the connected lines. So I looked for a pattern after making a table. This table shows that the number of total games increases by 1, 2, 3, 4, ⋯, when the number of classes increases one by one.
>
> Teacher: Min-su also did a good job. What do you think of this?
>
> Jun-ha: It is a good idea but you have to find out several cases to discover the pattern in a table.
>
> Min-su: I agree with Jun-ha. I would like to know another solution method except drawing a picture or making a table.
>
> Jun-ha: ⓐ We can write an equation by using the fact that each class has to play with the other classes except themselves.
>
> Teacher: Jae-hi, Min-su, and Jun-ha all did a great job. The following problem is similar to what you have solved. Solve this problem using the best method you can think of.
>
> <Problem>
>
> ⓑ *Every person attending a meeting shakes hands with all participants once except herself. If the number of total handshakes is 55, figure out the number of total participants.*
>
> 1) Write the equation of ⓐ, and solve the problem ⓑ using the same method [1 point].
>
> 2) The mathematics curriculum according to the revised national curriculum in 2009 describes four cautionary notes related to teaching and learning methods for developing mathematical creativity. Describe two cautionary notes with the cases

	Cautionary notes	Case implemented
Example	Stimulate students' divergent thinking by mathematical tasks producing various ideas.	The teacher provided students with problems which could be solved in many methods.
①		
②		

implemented in the lesson above except one example written for you [2 points].

Figure 1-1. Sample written-test items of the NTET

In the above sample item, the first part asks pre-service teachers to solve the given mathematics problem. Through these types of items, the NTET assesses whether pre-service elementary teachers have a basic mathematical knowledge. About two to four out of 11 points are allotted for this type of item.

The second part of the sample item assesses whether pre-service teachers understand a recent change of the national mathematics curriculum that emphasizes mathematical creativity[14]. In the most recently revised version of the national curriculum, mathematical creativity, which was mainly emphasized for gifted students, is also emphasized for teaching all students[21]. Given this, it is important for a teacher to understand instructional methods to foster such creativity through an ordinary mathematics lesson. It is noteworthy that simply knowing such methods written in the curriculum is not enough. Pre-service teachers are further expected to understand in what ways such methods could be implemented in a lesson.

4.3 *The In-depth Interview and Performance Test of the NTET*

After passing the written test of the NTET, pre-service elementary teachers need to take the second part of the NTET. This usually includes an in-depth interview, a lesson plan and implementation, and a teaching demonstration of an English lesson. Whereas the written test, consigned by the Korean Institute of Curriculum and Evaluation, is the same across the country, the details of the second test slightly differ across provinces.

For instance, Figure 1-2 shows some features of the second test of the NTET administered by the Seoul Metropolitan Office of Education.

Subject	Score	Test Range
In-depth interview	40	Aptitude as a teacher, view of teaching, personality & knowledge
Lesson plan	10	Constructing a lesson plan of the given topic
Lesson implementation	35	Communication skills and teaching ability
English lesson and interview	15	Conducting an English lesson and communicating in English

Figure 1- 2. Some features of the second test in the NTET

As shown above, about half of the total scores are assigned to lesson plan and lesson implementation. During the test, pre-service teachers write a full lesson plan on a given topic of a given subject matter for an hour. Even though the subject matter for writing a lesson plan varies across provinces and from year to year, writing a lesson plan for teaching mathematics is frequently assessed. The criteria for evaluating a lesson plan include the adequateness of lesson objectives, content organization, main activities, and instructional methods.

The performance test has been administered differently from year to year, especially in Seoul. For instance, pre-service teachers were asked to deliver a lesson plan they had constructed on the previous day in the 2013 NTET, whereas they had to design another lesson for the performance test on the spot in a relatively short period of time, and then to give a brief teaching demonstration based on the lesson plan for 15 minutes in the 2014 NTET. Regardless of the format of the test, however, it is clear that the NTET calls for a competent performance of teaching subject matters. The criteria for evaluating teaching performance include making a connection to the previous lessons, giving clear guidance of lesson objectives for students, developing main activities in a systematic and coherent way, adopting appropriate teaching methods, and skillfully using instructional materials.

The recent NTETs have put an emphasis on the performance test more than on the written test. Because of the importance of skillful teaching in the final process of the NTET, pre-service elementary teachers are expected to foster their teaching performance during the teacher preparation period.

5   Issues and Challenges in Elementary Teacher Education Programs

Given that the competencies of teachers have a significant impact on students' learning, there has been increased interest in teacher education programs[3,13]. In Korea, once employed as a public elementary school teacher, the job is secured until the retirement age of 62 without necessarily requiring further qualification. Even though in-service teachers are required to take several courses, training, and professional development programs, they are not necessarily required to take courses on teaching mathematics[10]. For this reason, high quality is required even for beginning teachers.

As described above, because of the high entry selectivity of elementary teacher education program and the final filtering process via the NTET, the overall quality of elementary school teachers in Korea is high. Most pre-service teachers feel confident with their readiness to teach elementary school mathematics in general, given their training and experience provided by teacher education programs[12].

However, there are issues and challenges that elementary teacher education programs in Korea face. The first issue relates to the connection between theory and practice, or between coursework offered by institution (e.g., what is taught in the university) and the needs of elementary schools (e.g., what is actually needed in teaching elementary students). For instance, Shin and Oh[25] surveyed pre-service and in-service teachers about the quality of elementary mathematics education programs in terms of program fulfillment, curriculum organization, facilities, and learning opportunities. They found that most teachers considered the fulfillment of the mathematics education programs as above average, but they called for a more close connection between theory and practice. In a similar vein, pre-service teachers often want to learn more about specific instructional methods or how to apply what they learned in the university to their actual teaching practice[11].

To address the gap between theory and practice, several efforts have been made in Korea. For example, mathematics methods courses include video-cases in order to show exemplary elementary mathematics instruction or to provoke discussion about how to effectively teach mathematics to students[19,20]. In another example, mathematics methods courses provide more opportunities to develop a lesson plan, to give a teaching demonstration, and to reflect on a lesson[16,24]. Another meaningful change is the strengthening of student teaching[7]. In fact, most universities have increased the minimum requirement of the fieldwork experience so that pre-service teachers have more opportunities to learn about teaching subject matters and classroom management before they will be solely responsible as a classroom teacher.

The second issue is related to subject matter emphasis. As elementary school teachers are expected to teach multiple subjects in Korea, they are required to take various courses dealing with knowledge and skills related to each subject matter during their teacher education programs. On the one hand, the strong academic background in multiple subjects enables pre-service teachers to adopt an inter-disciplinary approach and to gain sophisticated understanding of pedagogical knowledge applicable to any subject matter.

On the other hand, it is difficult for pre-service teachers to develop a profound understanding of each subject matter. For example, as described above, most pre-service teachers are required to take at least one course dealing with the foundations of pure mathematics, and then to take two more courses dealing with general theories and instructional methods related to teaching elementary mathematics. However, it has often been pointed out that pre-service teachers do not receive enough mathematical training for their future teaching careers[9,12,25]. This challenge is particularly true for mathematics or Korean language, taught more hours per week than any other subjects. Pre-service teachers should have more opportunities to learn content and methods related to these subjects, but this is difficult partly because of the structure of teacher education programs.

An alternative approach might be that pre-service teachers could develop profound knowledge in at least their concentrations. For instance, it is expected for pre-service teachers with mathematics concentration to have strong backgrounds both in pure mathematics and in mathematics education by taking about 21 credit hours. In fact, Pang[19] demonstrated that pre-service teachers with mathematics concentration, who at first focused more on the general features of a lesson which can be common across multiple subject matters, were able to notice the substantive characteristics of a *mathematics* lesson while taking a course of mathematics concentration programs.

As there has been lack of research on Korean elementary teacher education programs in international contexts, this chapter contributes to provoking a discussion on the nature and development of teacher education across different education systems. In particular, the issues and challenges may broaden our view on teacher education programs nested within unique cultural contexts.

References

1. Ball, D. L. & Forzani, F. M. (2009). The work of teaching and the challenge of teacher education. *Journal of Teacher Education, 60*, 497-511.
2. Choi, C. (2014). Study of teaching materials and methods for elementary mathematics. Seoul, Korea: Kyungmoon. [in Korean]
3. Feiman-Nemser, S. (2001). From preparation to practice: Designing a continuum to strengthen and sustain teaching. *Teachers College Record, 103*(6), 1013-1055.
4. Kang, J-H. (2009). Selecting the syllabus of mathematics as a general education course in national university of education. *Journal of Gongju Education, 46*(2), 1-24. [in Korean]
5. Kang, M., Kang, H., Kwon, S., Kim, S., Nam, J., Park, K. et al. (2013). *Understanding elementary mathematics education* (3rd Ed.). Seoul, Korea: Kyungmoon. [in Korean]
6. Kim, H. B. (2008). Factors of competence to be a 'classroom-friendly' teacher. *Korean Journal of Teacher Education, 24*(3), 4-15. [in Korean]
7. Kim, R. Y., Ham, S-H., & Paine, L. W. (2011). Knowledge expectations in mathematics teacher preparation programs in Korea and the United States: Towards international dialogue. *Journal of Teacher Education, 62*(1), 48-61.
8. Kwon, O. N, & Ju, M-K. (2012). Standards for professionalization of mathematics teachers: Policy, curricula, and national teacher employment test in Korea. *ZDM The International Journal on Mathematics Education, 44*, 211-222.

9. Lee, D-H. (2013). Applying Lakatos methods to the elementary pre-service teacher education. *Journal of Educational Research in Mathematics, 23*(4), 553-565. [in Korean]

10. Lee, H. C. (2012). *Inquiry of policy by analyzing teacher training courses in mathematics education.* Seoul, Korea: Korea Foundation for the Advancement of Science and Creativity. [in Korean]

11. Lee, Y., Kwon, J., & Lee, B. (2013). Changes in teacher efficacy beliefs in mathematics of elementary pre-service teachers during student teaching. *Journal of Educational Research in Mathematics, 23*(4), 407-422. [in Korean]

12. Li, Ma, & Pang, J. (2008). Mathematical preparation of prospective elementary teachers: Practices in selected education systems in East Asia. In P. Sullivan & T. Wood (Eds.), *The international handbook of mathematics teacher education: Vol. 1 Knowledge and beliefs in mathematics teaching and teaching development* (pp. 37-62). Rotterdam, Netherlands: Sense.

13. Masingila, J. O., Olanoff, D. E., & Kwaka, D. K. (2012). Who teaches mathematics content courses for prospective elementary teachers in the United States? Results of a national survey. *Journal of Mathematics Teacher Education, 15*, 347-358.

14. Ministry of Education, Science, and Technology (2011). *Mathematics curriculum.* Seoul, Korea: Author.

15. Na, G. S. (2008). Examining the prospective elementary teachers' perspective on mathematics class: Focused on the comparison of the comments on the mathematics class. *School Mathematics, 10*(2), 279-296. [in Korean]

16. Na, G. S. (2010). A study on the construction of mathematical knowledge by elementary pre-service teachers. *School Mathematics, 12*(2), 151-176. [in Korean]

17. Nam, S., Lew, S., Kwon, S., Kim, N., Shin, J., Park, S. et al. (2012). *Theory for teaching elementary school mathematics under the 2009 revised curriculum.* Seoul, Korea: Kyungmoon. [in Korean]

18. Organization for Economic Co-operation and Development (2007). *Education at a glance, 2007: OECD indicators.* Paris: Author.

19. Pang, J. (2011a). Case-based pedagogy for prospective teachers to learn how to teach elementary mathematics in Korea. *ZDM The International Journal on Mathematics Education, 43*, 777-789.

20. Pang, J. (2011b). Prospective teachers' analysis and conception of elementary mathematics instruction. *Journal of Elementary Mathematics Education in Korea, 15*(2), 221-246. [in Korean]

21. Pang, J. (2014). Changes to the Korean mathematics curriculum: Expectations and challenges. In Y. Li & G. Lappan (Eds.), *Mathematics curriculum in school education* (pp. 261-277). New York: Springer.

22. Park, K. (2010). Mathematics teacher education in Korea. In F. K. S. Leung & Y. Li (Eds.), *Reforms and issues in school mathematics in East Asia: Sharing and understanding mathematics education policies and practices* (pp. 181-196). Rotterdam, Netherlands: Sense.

23. Reys, R. E., Lindquist, M. M., Lambdin, D. V., & Smith, N. L. (2009). *Helping children learn mathematics* (9th Ed.). Hoboken, NJ: John Wiley & Sons. Park, S.,

Kim, M., Pang, J., & Kwon, J. (Translated into Korean) (2012). Seoul, Korea: Kyungmoon.

24. Seo, D. Y. (2010). A comparative study between the lectures on the practices of mathematics education in the courses for pre-service elementary teachers of two universities in the United States and Korea: Focused on two professors' cases. *Journal of Elementary Mathematics Education in Korea, 14*(3), 547-565. [in Korean]

25. Shin, H., & Oh, Y. (2005). An analysis on mathematics education programs of national universities of education. *Korean Elementary Education, 16*(1), 81-108. [in Korean]

CHAPTER 2

THE PIPELINE TO BECOMING AN ELEMENTARY TEACHER AND OPPORTUNITIES TO LEARN MATHEMATICS FOR TEACHING IN KOREA

Yeon Kim

Mathematics Education,
University of Michigan
610 East University Ave., Ann Arbor, MI 48109, USA
E-mail: yeonkim10@gmail.com

This chapter introduces the pipeline to becoming an elementary teacher in Korea, from entry through the teacher education program to getting advanced teacher certification. It also explains curriculum of teacher education programs and professional development. This chapter also clarifies the opportunities teachers have when it comes to learning about teaching mathematics. This chapter closes with some issues to offer enough opportunities for all teacher candidates to develop their mathematical knowledge and skills for teaching.

1 Introduction

Researchers have emphasized the importance of teacher quality in improving student achievement and in promoting, ultimately, a nation's economic competitiveness[2,3]. Many questions still remain, however, about the highly qualified teachers. For example, who are highly qualified teachers? To be highly qualified, what kind of knowledge and skills are required and what kind of teacher education establishes them. Irrespective of discussions, one thing that is critical for teaching quality is educational opportunities. To explain what educational opportunities are enjoyed by elementary teachers in Korea, this chapter provides an overview of (i) pathways to be a teacher in elementary school and (ii) opportunities to learn mathematics for teaching. Because mathematics is one of ten subjects that elementary teachers teach in Korea, the current chapter specifies the overall pipeline to becoming a teacher and in each step clarifies which courses teacher candidates can take to develop their knowledge for teaching mathematics.

2   Centralized System of Teacher Education and Certification

In Korea, nearly all aspects of teacher education programs and certification process are governed by the Ministry of Education. For elementary school levels, most teacher candidates are, as shown in Table 2-1, trained at the national universities of education. These universities

Table 2-1. Number of Incoming Students in Teacher Education Programs for Elementary School

Teacher education programs for elementary school	Year			
	1999	2004	2009	2014
Busan National University of Education	360	613	479	390
Cheongju National University of Education	350	463	400	314
Chuncheon National University of Education	400	538	450	350
Daegu National University of Education	520	614	513	422
Gyeongin National University of Education	560	640	801	658
Gongju National University of Education	470	574	481	381
Gwangju National University of Education	360	520	435	361
Jeonju National University of Education	300	443	383	306
Jeju National University of Education[a]	120	160	·	·
Jinju National University of Education	405	540	452	351
Seoul National University of Education	440	510	479	394
Jeju National University Sara Campus	·	·	134	119
Department of Primary Education at Korea National University of Education	160	160	135	128
Department of Primary Education at Ewha Womans University	50	50	40	39
Total	4495	5825	5182	4213

Source: Website of the Ministry of Education in Korea, www.mest.go.kr

---

[a] It merged with Cheju National University and renamed Sara Campus of Cheju National University in 2008.

offer only teacher education programs for elementary schools. Except for the Ewha Womans University, all universities were founded by the national government. The demand and supply of elementary school teachers are appropriately matched to maintain a purpose-oriented, teacher-training system[5]. This allows for the majority of teacher candidates for elementary school to get teaching positions. Korea's birth rate has been decreasing since the middle of 2000s, so a shortage in demand is expected to produce an over-supply of teachers.

The teacher education programs for elementary school have been transformed according to social and educational circumstances[5]. A lack of teachers prior to the 1980s compelled the government to implement various policies to increase their numbers. Between 1945 and 1961, one of Korea's urgent issues was securing elementary school teachers; hence the courses to become one were as short as three months and no longer than one year. Between 1961 and 1980, Korea established two-year teacher education programs at Universities of Education. The Ministry of Education broadened their focus to not just securing elementary school teachers but also monitoring their qualifications. Since the 1980s, policies for teacher education programs have focused on improving the qualification of teachers. In 1981, all Universities of Education began to offer four-year programs and continue to do so now. After completion of the current programs, all graduates receive a bachelor's degree in elementary education and the initial teacher's certificate is awarded without any requirement for a licensure examination. This certificate is valid for life.

3   Desirability of Teaching Profession

In Korea a teaching position is very desirable. This is true for several reasons. According to Kwon[6], Korean society still places a higher social status on teachers, a position not unlike, in Confucian tradition, the levels of king and father. The general public in Korea believes teachers make a greater contribution to society than do any other profession[1]. Once teaching became a high-status profession, more talented people become teachers, lifting the status of the profession even higher. Barber and Mourshed[1] claim that this feedback loop works well in Korea.

More current and critical reason is job stability. Since the financial and economic crisis of the late 1900s, a teaching position is regarded as being more stable than other jobs. Unless they have reason to be fired, teachers may work until they are 62, the retirement age. Another reason would be the economic benefits[6]. Most elementary school teachers are government civil servants, and they receive their pension from the twentieth year of work.[b] Furthermore, they are also eligible for better pension schemes, which is a kind of financial institute for teachers. In fact, teacher salaries in Korea are among the highest in the world. According to OECD[8], salaries for elementary teachers in Korea can rise to more than two and half times the starting salary. It is higher than the average maximum teacher salary in the OECD as shown in Figure 2-1. The corresponding figures of the United States at that time were less than two times the starting salary.

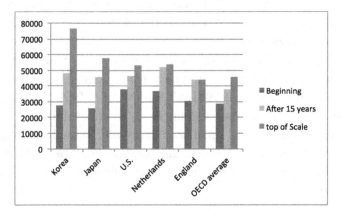

Note: Salaries are converted to U.S. dollar equivalents using a purchasing power parity index.

Source: OECD, Education at a glance: OECD indicators 2013, Paris, 2013

Figure 2-1. Annual Salaries for Elementary Education Teachers in U.S. Dollars, 2011

---

[b] In 2011, there are 5,789 public elementary schools and there are 75 private elementary schools in South Korea. Thus, most elementary school teachers are government civil servants.

4 Entry Requirements into Teacher Education Programs and Who Enters the Universities of Education

In Korea, entry requirements for teacher education programs are based on students' senior high school records (achievement level in each subject area and homeroom teacher's recommendation) and their performance on the College Scholastic Ability Test (CSAT).[c] Teacher education programs also interview applicants about teaching attitudes and ethics, which make up 5-10% of an applicant's overall score. Each teacher education program can determine its own evaluation criteria.

Entry to teacher education programs in Korea is difficult and competitive. Few spots are available and they are often allocated through the high-level and high-stakes upper secondary examination. Since entry to teacher education programs usually requires university student status, the relative difficulty of university entry is pertinent to any comparison of the rigor of teacher education program. In fact, teacher education programs recruit their incoming students in the top five percent from each cohort that graduate from their school systems in Korea, the top ten percent in Finland, and the top 20 percent in Singapore[1,4]. As reported in TIMSS study[7], students in Korea are always more highly ranked in mathematics. Although there is no comparison study to show the status of incoming students in teacher education programs, Korean applicants have very strong mathematics subject knowledge. Because of this, instructors in teacher education programs for elementary schools generally do not face the challenge of overcoming the teacher candidates' lack of mathematics knowledge.

5 Curriculum in Teacher Education

As Yoon, Song, Jo, and Kim[11] pointed out, teacher education programs for elementary school have similar systems, organization, and curriculum. Each designs and implements a curriculum under Articles 41 and 44 of the Higher Education Act in accordance with the relevant rules set by the

---

[c] This test is like the Scholastic Assessment Test (SAT) in the United States.

Ministry of Education.[d] The curriculum in teacher education programs for elementary school consists of five parts: (i) courses for the liberal arts, (ii) courses for studies of education, (iii) courses for subject matter education, (iv) courses for advanced subject knowledge including thesis, and (v) field experience. For an efficient explanation, this section uses the curriculum of Seoul National University of Education to specify course titles and numbers of courses and credits. There are some differences among all teacher education programs for elementary school, but, again, they have very similar course titles and credits.

5.1 *Courses for the Liberal Arts*

Courses for the liberal arts include 20 courses for 40 credits offering introductions or foundations of philosophy, literature, language, history, social science, mathematics, and natural science, such as Language and Culture, Understanding of Philosophy, Writing, Introduction of Natural Science, English Conversation, Understanding of Information Science, Understanding of Literature, etc. Generally, freshmen and sophomores take these courses.

There are two courses related to mathematics: Foundation for Modern Mathematics is a required course, and Essence of Mathematics is an elective.

5.2 *Courses for Educational Theory*

Educational theory deals with understanding children, guidance, educational phenomenon and execution of elementary education[5]. There are eight required courses: Historical and Philosophical Foundation of Education, Understanding of Special Children, Curriculum and a Lesson,

---

[d] "Higher Education Act" Article 41: (1) The purpose of the university of education is to foster teachers for elementary schools. "Higher Education Act" Article 44: The education in the universities of education, colleges of education, comprehensive universities and colleges of education and the departments of education shall be conducted for students to fulfill any of the following purposes to realize the purposes of the establishment: 1. Be sure to have firm value and sound ethics as instructors; 2. Be sure to experience the education philosophy and specific practical method; 3. Be sure to establish a basis to increase quality and capability of instructor for his/her whole life.

Educational Psychology and Counsel, Education and Society, Prevention and Countermeasure of School Violence, and School and Class Management. Teacher candidates select one among three courses: Understanding of Early Children and Elementary Education, Educational Technology, and Educational Evaluation and Measurement. Finally, teacher candidates take 18 credits in the courses for educational theories.

5.3 *Courses for Subject Matter Education*

Courses for subject matter education deal with ten subjects (morality, Korean language, social science, mathematics, natural science, physical education, music, art, English, and practical course education), integrated subjects, and extracurricular subjects. This is because elementary school teachers teach those subjects in Korea. There are two types of courses for each subject: one concerns the theoretical approach and the other the practical approach. Thus, the former one includes the contents course, generally called worldly, and educational studies about each subject; the latter one includes a kind of methods course. In total, there are 58 credits. Korea uses a national curriculum for grades 1-9, and in particular all elementary school students and teachers use the same textbooks and teacher guides developed by the government. Because of this, it is very clear for instructors and teacher candidates to figure out what contents should be taught in elementary school and how those contents are represented in textbooks that elementary school teachers and students use.

Among them, there are two courses related to mathematics: Understanding of Elementary Mathematics Education with two credits and Practice of Elementary Mathematics Education with three credits.

5.4 *Courses for Advanced Subject Knowledge*

Courses for advanced subject knowledge are designed to help teacher candidates study particular subject matter knowledge in depth. Because there are ten subjects, around 10% teacher candidates take courses for advanced subject knowledge in each subject. For example, there are four required courses for mathematics: Elementary Mathematics Problem Solving, Theory In Elementary Mathematics Education, Research

Methodology In Elementary Mathematics Education, And Curriculum And Evaluation In Elementary Mathematics. There are around 15 selective courses, such as Algebra and Elementary Mathematics Education, Analysis and Elementary Mathematics Education, Statistics And Elementary Mathematics Education, Geometry and Elementary Mathematics Education, Mathematical Heuristics, History of Mathematics And Mathematics Education, Mathematics Education for Gifted and Underachieved Students, Mathematics Education for Early Children, Study About Number and Operation, Study About Geometry And Measurement, Study About Pattern and Statistics, Game for Mathematics, Etc. The courses include nearly all content areas of mathematics for elementary school. Teacher candidates take around 20 credits.

5.5 Field Experience

According to Santagata, Zannoni, and Stigler (2007), in offering field experiences teacher education programs make two assumptions: (i) exposure to examples of teaching creates learning opportunities for teacher candidates; and (ii) through field experience teacher candidates meld theory into practice. In the 1980s, teacher candidates are required to complete at least a four-week field experience program at affiliated elementary schools. Since the 1990s, each teacher education program includes in its curriculum field experience programs lasting eight weeks, sending their teacher candidates to affiliated elementary schools or partner schools.

Field experience is conducted in various ways. Currently, some programs implement it once a semester for juniors and seniors while other programs include sophomores as well. There are various types of field experience. For example, in Seoul National University of Education, sophomores do one-week observations and two-week participations, juniors teach for four weeks, and seniors do practical training for two weeks.

Partner schools for field experience are selected by districts. Teacher education programs distribute their teacher candidates to each school. Field experience is one independent course. In other words, during field experience, coaching teachers in affiliated schools or partner schools guide teacher candidates without any cooperation from teacher education programs. Instructors in teacher education programs also cannot use teacher candidates' field experience. Furthermore, there is no one to manage what teacher candidates learn in class and experience in the field.

5.6 *What Learning Opportunities Teacher Candidates Have about Mathematical Knowledge Entailed in Teaching*

All teacher candidates must take Foundation for Modern Mathematics (two credits), Understanding of Elementary Mathematics Education (two credits) and Practice of Elementary Mathematics Education (three credits). Some can take as an elective, Essence of Mathematic*s* (two credits). Approximately 10% teacher candidates should take ten courses with 20 credits for advanced subject knowledge related to teaching mathematics. Related to mathematics and teaching mathematics, those teachers can take 14 courses with 29 credits in total, but the other teachers can take at best three courses with seven credits.

Committee on the Study of Teacher Preparation Programs in the United States[3] recommended "Prospective elementary grade teachers (teacher candidate) should be required to take at least 9 semester-hours on fundamental ideas of elementary school mathematics" (p. 115). According to this recommendation, 10% of teacher candidates in Korea exceed the learning opportunities related to mathematics for teaching, but the others have fewer opportunities. The number of courses and credits quantitatively show learning opportunities that teacher candidates can have. This committee also provides specific suggestions for teacher education program qualitatively:

The field of mathematics education has established a firm consensus that to prepare effective K-12 mathematics teachers, a program should provide prospective teachers (teacher candidates) with the knowledge and skills described by the Conference Board of the Mathematical Sciences:

a deep understanding of the mathematics they will teach,
courses that focus on a thorough development of basic mathematical ideas, and

courses that develop careful reasoning and mathematical "common sense" in analyzing conceptual relationships and solving problems, and courses that develop the habits of mind of a mathematical thinker (p. 117).

The curriculum of teacher education programs shows the overall structure and titles of courses, but what is hidden is the curriculum for each course. Hence, it is hard to figure out the purpose of each course is, the content discussed in each course, or how the purposes and contents are managed. In other words, it is unclear whether the courses related to mathematics or teaching mathematics in Korea qualitatively help teacher candidate have enough knowledge and skills entailed in the teaching of mathematics or, if they do, how. As shown in the previous sections, teacher candidates in teacher education programs have very good content knowledge in mathematics, and teacher education programs in Korea offer lots of courses for ten subjects and for educational theory. Teacher education programs might expect that teacher candidates can synthetically comprehend their learning from courses that they have taken and apply it in their actual classrooms later. A promising area of research in Korea is to scrutinize the curriculums of each course in teacher education programs in terms of mathematics education for teacher candidates.

6   Employment Examination

To get a teaching position in elementary school, teacher candidates, who have already graduated or are candidates for graduation, must take and pass the government-administered Teacher Employment Examination.[e] Again, to graduate from a teacher education program grants one an initial teacher certificate; only certificate holders are eligible to take the employment examination.

This examination is composed of a multiple-choice test and an essay test for the first round and an interview and performance assessment for the second round. To advance to the second round, a candidate must pass the first round. The first round focuses on educational theory and the national curriculum for all ten subjects in elementary school. The applicant also needs to demonstrate teacher aptitude. In the second round, the applicant must (i) demonstrate his or her communication skills as a teacher, (ii) design a lesson plan on a certain chapter in the curriculum, and (iii) demonstrate English communication skill.

Teacher candidates' knowledge and skills related to teaching mathematics are examined in the all rounds with regards to understanding of mathematics curriculum for elementary school and writing a lesson plan for teaching mathematics. Because this examination elicits such intense competition, seniors in teacher education programs prepare arduously for this examination. In fact, Wang, Coleman, Coley, and Phelps (2003) considered this examination as a high-stakes filter in the teacher development pipeline.

7 Professional Development for Elementary Teachers

Korea offers various types of professional development. The most important one is the 180 hours of training at programs approved by the Ministry of Education for advance certification. After three years of teaching, teachers are eligible to take this professional development during their winter or summer break. The successful completion gives

---

[e] Again, there are small numbers of private elementary schools in South Korea, and teacher candidates in teacher education programs for elementary schools generally would like to get teaching positions in public schools. Thus, being an elementary school teacher is generally known as being a government civil servant after passing the employment examination by the metropolitan and provincial offices of education.

teachers advance certification with a better chance for later promotion to head teacher, an administrative position with a small salary increase.

Most teachers go through with this professional development. Although it could be an opportunity for teachers to improve their knowledge and skills for teaching mathematics, its curriculum is not generally regarded as such. There is no connection between the curriculum of teacher education programs and that of this professional development. Furthermore, as Korea has no evaluation for tenure, this professional development is a final chance to evaluate teachers.

8   Closing Remarks

Korea has a very centralized system for teacher education and certification. Thirteen universities have teacher education programs for elementary school, and ten of them are universities of education, which are only for teacher candidates for elementary school. Although each teacher education program can design and implement a curriculum under the Higher Education Act in accordance with the rules set by the Ministry of Education, universities of education use very similar curriculum. Most teacher candidates take three courses for seven credits about the theory and practice of mathematics education for elementary schools. However, approximately 10% of teacher candidates take 14 courses for a total of 29 credits. It would be important, on one hand, to provide enough opportunities for all teacher candidates to develop their mathematical knowledge and skills for teaching. On the other hand, it would be critical to gather and share ideas and cases about what curriculum can be designed and implemented in courses of mathematics education for teacher candidates. Furthermore, as a long-term policy, it could prove beneficial to connect the curriculums of teacher education programs for the initial teacher certificate and professional development for advanced certificate. Doing so could help teachers have continuous learning as teachers and improve their knowledge and skills for teaching mathematics.

References

1. Barber, M., & Mourshed, M. (2007). *How the best performing school systems come out on top.* New York, NY: McKinsey & Company.
2. Cohen-Vogel, L. (2005). Federal role in teacher quality: "Redefinition" or policy alignment? *Educational Policy, 19*(1), 18-43.
3. Committee on the Study of Teacher Preparation Programs in the United States. (2010). *Preparing teachers: Building evidence for sound policy.* Washington, DC: The National Academies Press.
4. Ingvarson, L., Schwille, J., Tatto, M.T., Rowley, G., Peck, R., & Senk, S.L. (2013). *An analysis of teacher education context, structure, and quality-assurance arrangements in TEDS-M countries: Findings from the IEA Teacher Education and Development Study in Mathematics (TEDS-M).* Amsterdam: IEA.
5. Kim, K., Chung, M., & Kim, D. (2012). *Successful strategy for training teachers in Korean education.* Seoul: Ministry of Strategy and Finance.
6. Kwon, O. (2004). *Mathematics teacher education in Korea.* Paper presented at the International Congress on mathematical Education (ICME-10), Copenhagen, Denmark.
7. Mullis, I. V. S., Martin, M. O., Foy, P., & Arora, A. (2012). *TIMSS 2011 international results in mathematics.* Chestnut Hill, MA: TIMSS & PIRLS International Study Center.
8. OECD. (2013). *Education at a glance 2013: OECD indicators.* Paris, France: OECD Publishing.
9. Santagata, R., Zannoni, C., & Stigler, J. W. (2007). The role of lesson analysis in pre-service teacher education: an empirical investigation of teacher learning from a virtual video-based field experience. *Journal of Mathematics Teacher Education, 10*(2), 123-140.
10. Wang, A., Coleman, A. B., Coley, R. J., & Phelps, R. P. (2003). *Preparing teachers around the world.* princeton, NJ: Educational Testing Service.
11. Yoon, J., Song, K., Jo, D., & Kim, B. (1997). *Korea education policy study.* Seoul: Koyookkwahacksa. [in Korean]

CHAPTER 3

STUDENT TEACHING IN MATHEMATICS EDUCATION AT THE SECONDARY LEVEL IN KOREA

Chonghee Lee

Department of Mathematics Education
Ewha Womans University
52, Ewhayeodae-gil, Seodaemun-gu, Seoul, South Korea
E-mail: jonghee@ewha.ac.kr

Hwa Young Lee

Department of Mathematics and Science Education
University of Georgia
105 Aderhold Hall, Athens, GA 30605
E-mail:hylee00@uga.edu

In this chapter we introduce the student teaching system in mathematics education at the secondary level in Korea. First, we introduce the background information that is needed to understand the student teaching system in Korea. Second, we articulate the definition and goals of student teaching. Third, we explain the participants of the student teaching process and their roles. Then we provide the timeline, main activities, and products throughout student teaching required at Ewha Womans University[a]. We close with a discussion on how the student teachers are evaluated and present research on the perception of student teaching.

1 Introduction

In the process of preparing secondary mathematics teachers, it is important for teacher candidates to have experiences to connect educational theory with experiences in secondary schools. Research has shown that student teaching is an effective and necessary process in the

---

[a] Ewha Womans University is a private female university located in the capital city, Seoul. We use the secondary student teaching program at this institute as an illustrative example.

preparation of prospective mathematics teachers[1,7,9]. In this chapter, we introduce the student teaching process that is required for prospective secondary mathematics teachers in Korea.

In order to become a mathematics teacher at the secondary level, there are several ways to obtain certification. At the undergraduate level, students in a four year college program who major in mathematics education or who major in a related field and meet the alternative requirements can obtain certification of teaching secondary mathematics. At the graduate level, students who complete a secondary teacher preparation program in mathematics education are certified to teach secondary mathematics. These teacher candidates will be licensed to teach at middle or high schools in Korea from grades seven through twelve after completion of the aforementioned teacher preparation programs. After being certified, they can either apply to work at private schools or take the public school system entrance examination administered by each city and become a public school teacher.

Each university that is accredited for granting teacher certification has a student teaching program which is mandatory for all prospective teachers. Student teaching programs vary slightly by each institute; Thus it is difficult to describe a generic system that entails all these different programs. However, research efforts have found some commonalities throughout these programs in defining student teaching and explaining the overarching goals of student teaching1,3,4,9. In the following section, we provide a definition of student teaching and synthesize some of the major goals of student teaching that have been identified in existing research.

2   Definition and Goals of Student Teaching

2.1 *What is Student Teaching?*

Shim and Lee[9] defined student teaching as a process of applying and putting into practice the educational theories learned at the teacher education institute through working in the actual educational field. Student teaching is also a process of evaluating, executing, and testing ones skills and aptitude towards teaching. Lee and Sohn[3] categorized four approaches scholars take when understanding the concept of student teaching. The first is student teaching as a mandatory process for obtaining teacher certification. Second, student teaching can be considered as practicing educational theories in the field: It is an opportunity to apply and reflect on theory based on experience in schools and classrooms. The third approach is to consider student teaching as a chance to reflect on oneself as a teacher: It is a learning experience to check one's aptitude towards teaching and helps making a decision whether or not to enter the teaching profession. Lastly, student teaching is another way of learning as an experience: Through experiential learning in school settings, the student teacher learns about educational activities and teaching.

2.2 *Goals of Student Teaching*

In this section, we present some of the goals of student teaching in light of the approaches identified by Lee and Sohn[3]. These goals serve different purposes and act as the foundation of student teaching as a systematic whole. We do not consider any single goal more important than the others, but order the goals in correspondence with Lee and Sohn's[3] categorization of the approaches taken when understanding the concept of student teaching.

The first goal of student teaching is a practical one in that it is to obtain teacher certification. According to the twelfth and twentieth article of the enforcement regulations of the ordinance of teacher licensure in Korea (which can be downloaded at www.law.go.kr), student teaching is a requirement for teacher certification. According to this regulation, in order to obtain a secondary teacher certification teacher candidates have to complete at least four weeks of student teaching[3]. Thus, in order to be accredited as a public or private school teacher student teaching is a necessary process.

The second goal of student teaching is to apply and reflect on the educational theories learned throughout the teacher preparation program. This goal consists of two sub-goals. First, the purpose of student teaching is in having the opportunity to apply educational theories in schools and classrooms. Secondly, a goal of student teaching is to learn ways to refine the educational theories in order to better fit the specific circumstances and conditions that are given by the situated teaching experience. To achieve this goal, it is necessary for prospective teachers to analyze and critique educational theories. Thus, it becomes important to understand the local conditions and problems of education that are not fully discussed in theory and to fill in the gaps of learning experience that are not covered through the education provided from the teacher education institute. Ultimately, the goal is to develop flexible and creative future teachers.

The third goal of student teaching is to explore teaching as a profession and develop a sense of duty and a teaching philosophy. Through the experience of teaching in a real school setting, the prospective teacher may learn whether they find teaching as a job adequate for him or her. The teaching experience provides an opportunity to evaluate and reflect on oneself as a teacher and to find the strengths and weaknesses one has as a future teacher. It would be unfortunate not only for the teacher candidate but also the potential students of that teacher candidate if teaching is not the best fit for him or her. Thus the goal of student teaching is to test one's desire to teach and aptitude for teaching. Moreover, the teaching experience provides a context for developing a sense of duty as a teacher and establishing one's unique philosophy of teaching.

Fourth, the goal of student teaching is in gaining real experience in actual schools and classrooms as a continuum of learning to become a teacher. It is to submerge oneself in a real school context and pursue all the duties required as a mathematics teacher in a middle or high school. This includes not only teaching mathematics but also serving as a homeroom teacher and doing administrative work. Many teachers are assigned as a homeroom teacher in addition to their subject teaching duties which includes supervising students' school life such as cleaning classrooms and monitoring students' attire, personal belongings and hygiene. Additionally, all teachers are assigned to do administrative work such as managing the school time table, assisting with the school talent show and other events, managing the attendance of the entire school, completing paperwork for student transfers, etc. These are critical duties that are required of teachers that are difficult to learn without being placed in a school environment. Therefore the goal of student teaching is to obtain experience in these various jobs a teacher handles in the everyday teaching life as a continuum of experiential learning. This goal provides the teacher candidate an opportunity to find skills and knowledge to further develop and work on when preparing for becoming a teacher.

3 Participants and Their Roles

The student teacher, supervising professor from the institute, and the cooperating mentor teacher of the school form the triad of participants in the student teaching process. In the following sections we articulate the qualifications, roles, and duties of participants. These discussions are based on studies by other researchers[3,4] and our own experiences as a student teacher, mentor teacher, and university supervisor.

3.1 *The student teacher*

As Lee and Sohn[3] explained, the qualifications for a prospective teacher to pursue student teaching at the secondary level in Korea are as the following: In accordance with the fifteenth article of the enforcement regulations of teacher licensure, the qualifications are either one of the following (a) To be enrolled as a student in the college of education or graduate school of education and have no legal blemishes that would present obstacles in obtaining certification or (b) for those who are not enrolled in an education department at the four year college level, those who rank within the top 30% of students within any major field of study can apply for a course in teacher education within the first twenty days of the first semester of the second year. Among these students, those who were selected by the institute based on their personality, aptitude, and grades are reported to the Minister of Education & Human Resources Development within the first 60 days of the first semester of the third year. If the grade point average in the major subject and education courses are higher or equal to a B grade and there are no legal blemishes that would present obstacles in obtaining teacher certification then the student meets the qualification to pursue in student teaching.

Once the teacher candidate meets the aforementioned qualifications for student teaching, the student teacher is assigned to a school that is either an affiliated school ran by the same corporate body of the institute or a cooperating school that agreed to participate in the program. Usually student teachers are assigned individually to schools and mentor teachers but occasionally they are assigned in pairs. At the assigned school the student teacher has to learn how to balance various roles. These roles are learned and carried out through observations and practice. Usually the student teacher begins by observing these activities and gradually assumes more authority and responsibility as time progresses. The activities are guided by the student teaching handbook provided by the institution.

Roles associated with student teaching can be described generally in terms of three main categories: mathematics content, homeroom management, and other administrative duties. First, the student teacher is expected to teach the mathematical content as a mathematics teacher. Usually, the student teacher observes, participates, and gradually assumes responsibility for teaching the lessons for the classes assigned to the mentor teacher. The first phase of student teaching involves observing other teachers. According to Shim and Lee[3], the student teachers are expected to observe and experience the entire process of planning, preparing, executing, and reflecting on a mathematics lesson using a practicing teacher as an example. Based on these observations, the student teacher gradually assumes authority over instruction. The number of lessons that are required for student teachers to teach varies slightly by institute. In the case of Ewha Womans University, the student teacher is required to observe at least five lessons and teach at least five lessons. Also, the student teacher is required to prepare an instruction demonstration to be observed by the supervising professor, mentor teacher, and other student teachers. The student teacher is expected to submit a formal lesson plan for this lesson to the supervising professor.

Second, the student teacher is expected to learn how to manage a homeroom. In the Korean school system, every student in each grade level is assigned to one homeroom. Students within the same homeroom spend much of their time together at school: Subject teachers usually come to the homeroom classroom rather than the students moving from room to room; in some schools, students also eat lunch in their homeroom classrooms. Therefore, there is a homeroom teacher assigned to manage each homeroom. Each school expects slightly different things from the homeroom teacher but in general, the job of the homeroom teacher is somewhat like a "parent" of the homeroom. The job is to inform the students of upcoming events, share important information, and to otherwise supervise academic, extracurricular, and social life in school. These include supervising creative extracurricular activities, counseling, and supervising students' school life.

Third, the student teacher is expected to learn the administrative work that assigned to the mentor teacher. As mentioned earlier, each teacher in the school is in charge of one or more administrative duties in the school. As future teachers, this is a chance to experience some of the work that teachers do as an apprentice. Later in this chapter we provide examples of some of the duties that are expected for the student teachers in the student teaching program at Ewha Womans University.

Other duties for the student teachers are to collect assignments in their student teaching portfolio and keep a daily journal. Later in this chapter we provide examples of assignments from the student teaching program at Ewha Womans University. Usually student teachers are exempted from classes while they are student teaching. Thus, the main duties they are required for course work are to focus on student teaching and to maintain a student teaching portfolio. In the daily journal included in the student teaching portfolio, student teachers record the main activities they did each day and reflect on what they learned from them. There is a place for the mentor teacher to leave a comment as feedback in the journal. There are no other seminars or courses that are associated with the student teaching program. Instead, they are required to submit a lesson plan for the lesson that will be observed by the supervising professor from their institute and the student teaching portfolio at the end of student teaching. Typically there is one observation for each student teacher. The roles of the supervising professor will be further elaborated in the next section.

3.2 *Supervising Professor*

The process by which supervising professors are assigned varies slightly from institution to institution. Typically, the supervising professor from the institute is a professor from the related department and major. Supervisor responsibilities include observing a lesson and supervising the student teachers. According to Na et al.[4], the supervisor from the institute has the following roles and duties.

First, it is the supervising professor's job to cooperate with the school and plan ahead of time and set foundations for the student teaching process. This includes meeting with the teachers in charge of student teaching at the affiliated or cooperating schools and deciding the timeline and student teaching placements. The affiliated or cooperating school has to match the school level in which the student teacher is to be certified. Thus to be certified in the secondary level in mathematics, a student teacher has to be assigned to mentor teacher in either a middle school or high school. The supervising professor has to check whether the appropriate subject matter positions are available at these schools and determine the number of positions available. This information is critical for assigning the student teachers to their schools. Second, after the student teachers are placed at schools that meet these criteria, the supervising professor sets up an initial meeting with the student teachers. In this meeting, the professor explains the goals and importance of student teaching, and emphasizes the professional attitude, roles, and duties that are required from the student teachers. Also, in this meeting student teachers are informed of how to maintain their student teaching journals. Finally, in this meeting, the supervising professor collects the contact information of each student teacher for correspondence throughout student teaching. Third, throughout the student teaching process, it is the supervising professor's job to provide supervision and consultation as needed. The supervising professor either visits the school or talks with the student teachers online. Typically the supervising professor observes one lesson (called the instruction demonstration) for each student teacher. Prior to this observation, the student teacher submits a formal lesson plan to the supervising professor. Fourth, the supervising professor evaluates the student teachers and the student teaching process. The supervising professor assigns a grade to each student teacher based on the student teaching portfolio, lesson plan, class observation, and the evaluation from the mentor teacher. Lastly, the supervising professor sets up a final debriefing meeting with the teachers and administrators involved with the student teaching process to evaluate and reflect on the program.

3.3 *Mentor Teacher*

The mentor teacher is the teacher at the assigned school who takes part in educating the student teacher. As a future colleague in the teaching field, the mentor teacher is expected to serve as a role model for the student teachers and guide them while they are at the school throughout the four weeks of the student teaching[9]. The mentor teacher is usually decided within the mathematics department in the school at the beginning of the year. The number of years of teaching experience a mentor teacher has can vary by school and circumstances but it is typically a job given to teachers with more than three years of teaching experience. The mentor teacher who is assigned by the school to mentor the student teacher is ideally also a homeroom teacher. This allows the student teacher to have a single mentor teacher. However, if the mentor teacher who teaches mathematics does not have a homeroom class then the student teacher is assigned to a second mentor teacher who is in charge of managing homeroom. In this case the student teacher has a subject mentor teacher and a homeroom mentor teacher. Na et al.,[4] summarized the roles and duties of the mentor teacher as follows.

First, the mentor teacher supervises the observation process: the job of the mentor teacher is to oversee the classes the student teacher observes and how they observe the lessons. This includes seeking permission from other teachers so that the student teacher can observe their lessons and ensuring that student teachers behave professionally in others' classrooms. Second, mentor teachers guide and supervise student teachers in writing lesson plans, creating sequencing charts, and using appropriate resources, materials, and technology. Third, the mentor teacher observes the student teacher's teaching and classroom management and provides feedback. Fourth, the mentor teacher provides opportunities for the student teacher to apply educational theories into practice and improve teaching skills. Fifth, the mentor teacher supervises the preparation of the observation. This preparation includes advising the lesson planning and preparing the lesson and post-observation meeting. Sixth, the mentor teacher helps the student teacher become familiar with the school facilities, laboratories and learn how to manage them if needed. Seventh, the mentor teacher informs the student teacher about important administrative regulations and classroom or laboratory management rules. Eighth, the mentor teacher checks and provides feedback to the daily student teaching journals. Ninth, the mentor teacher supervises and provides opportunities to learn how to manage the homeroom. Homeroom tasks include morning and end of school day assemblies, classroom cleaning supervision, managing the classroom environment, managing the administrative work related to running the homeroom, supervising and counseling the students in the homeroom class.

4   Timeline and Main Activities

4.1 *When and for How Long?*

According to Na et al.[4], it is required that student teaching is completed for more than or equal to four credit hours including up to two credit hours of educational volunteering activities. Student teaching at a full day school counts as one credit hour per two weeks of placement. Thus student teaching is required for at least four weeks including observing, co-teaching, and teaching activities. Usually student teaching is completed in the senior year for four weeks. In some institutes student teaching placements are distributed throughout the years in the program. For example, student teaching can be completed one week in the sophomore year with a focus on observation, another week in the junior year with a focus on transitioning into teaching, and the last two weeks in the senior year with a focus on full-time teaching[4].

At Ewha Womans University, students who have completed or are planning to have completed two credit hours (60 hours) of educational volunteering by the second semester of their third year can apply for student teaching. Student teaching is done in their first semester of their fourth year in either April or May for four weeks.

4.2 *Timeline and Main Activities*

Na et al.[4] provided the general timeline that student teaching programs follow as depicted in Figure 3-1. In this figure, ST refers to student teacher and MT refers to the mentor teacher.

Stage	Main Activities		Participants
Preparation	Initial meeting and school assignment		Institute & ST
	Student teaching orientation at the institute		Institute & ST
	Visit assigned school		ST
	Student teaching orientation at the school		School & ST

↓

Execution	Observation		School & ST
	Participation (transitioning)		School & ST
	Practicing (teaching lessons and managing homeroom and administrative work)		School & ST

↓

Debriefing and Evaluation	At school	Putting student teaching results together	ST
		Self-evaluate student teaching	ST
		Evaluation of the student teacher from the school	MT
	At institute	Evaluation of the student teacher from the institute	Institute
		Evaluation of the program	Institute & School

Figure 3-1. Main Activities

In order to provide a more specific example of activities we introduce the guideline for the main activities recommended by the student teaching program at Ewha Womans University. Then we provide an example of the student teaching guidelines that were sent out to the affiliated middle school for student teaching in the fall of 2013.

First, Table 3-1 shows each weekly activity suggested in the Student Teaching Report and Refinement Plan. The results and implementation of this plan is reported in Seo, Joo, Lee, Hyun, Lee, Shim, and Kim[8]. In the table, '⊙'denotes the main focus activity required for the week and '•' denotes other additional activities that are required for the week. As shown in Table 3-1, the main activities start with collecting information about the school and observing lessons. Then the student teacher gradually takes over teaching lessons and managing the assigned homeroom class. By the fourth week, the student teacher is expected to manage administrative work as well. Throughout the student teaching, self-evaluation is an ongoing process. Table 3-2 is an example of the student teaching guidelines that were sent out to the affiliated middle school for student teaching in the fall of 2013. The specific assignments that are required for student teachers in the student teaching portfolio (which can be found at http://home.ewha.ac.kr/~ofe/) are provided in Table 3-3.

Table 3-1. Main Activities

Activity	Week 1	Week 2	Week 3	Week 4
Collect school information	⊙	•		
Observe lessons	⊙	•	•	•
Teach lessons			⊙	•
Manage homeroom			⊙	•
Supervise students			⊙	•
Manage administrative work			•	⊙
Evaluate	•	•	•	⊙

Table 3-2. Student Teaching Timeline and Weekly Activities Example

Activities	Week 1	Week 2	Week 3	Week 4
1. Subject and Homeroom assignment	→			
2. General supervision	→			
3. Subject matter supervision	→			
4. General lesson observations	→			
5. Observation of classes to teach		→		
1. Write lesson plans			→	
2. Teach lessons and post-observation meetings			→	
3. Create lesson material			→	
4. Supervise extracurricular activity				→
5. Supervise homeroom cleaning and lunch time				→
6. Manage morning and end of school day assembly for homeroom			→	
7. Manage homeroom administrative work				→
8. Instruction demonstration for each subject (assign one student teacher for each subject)			→	
1. Manage homeroom classroom environment		→		
2. Submit lesson material	Discuss	Plan	Complete	
3. Submit student teaching journal every Monday	Homeroom mentor teacher	Subject mentor teacher	Homeroom mentor teacher	Both mentor teachers
4. Evaluation meeting	Homeroom and Subject mentor teacher, final evaluation meeting			

Table 3-3. Weekly Assignments

Stage	Content	Activities
Observation	Collect School Information	1. Document the guidelines for student teaching at the school provided at the student teacher orientation. 2. Discuss the 'Student teaching guideline and assignment handbook' with mentor teacher and record the assignments or activities that need to be revised or changed due to the school's condition. 3. Research the school and record the history and educational goal and the school faculty directory 4. Record the monthly events of the school. 5. Write the weekly time table for the assigned homeroom class and the mentor teacher's mathematics classes.
	Observation	6. Observe students in the assigned homeroom class and record characteristics of each student. 7. Observe at least five lessons taught by mentor teacher or other teachers and record each lesson in the observation section. 8. Write a reflection after completing all observations on what was learned from the observations.
Practice	Teach Lessons	9. Write lesson plans for at least five lessons. Provide an overview of the lesson following the format in the student teaching portfolio and attach detailed lesson plan. If the school requires a certain format for the lesson plan follow that format and attach to the overview in the student teaching portfolio. 10. After each lesson, evaluate the lesson with mentor teacher and reflect on the things you learned from teaching the lesson. It is required to teach at least five lessons. 11. After the teaching is complete, reflect on the things you learned.

	Manage Homeroom	12. List the things you did in terms of managing the assigned homeroom class in the following activities: ⓐ Manage Student Roll ⓑ Manage classroom environment and cleaning ⓒ Supervise lunch time ⓓ Supervise students in charge of chores 13. In completion of managing the assigned homeroom class, reflect on the experience and write what you learned from it.
	Supervise Students' School Life	14. List the things you did in terms of supervising students' school life in the following activities: ⓐ Supervising general school life supervision ⓑ Career counseling ⓒ Counseling 15. In completion of supervising students' school life, reflect on the experience and write what you learned from it.
	Management	16. List the administrative work you practiced such as: ⓐ Writing official documents ⓑ Recording and documenting schools' affairs
Evaluation	Evaluating Student Teaching	17. Self-evaluate the student teaching experience over the four weeks. Write about what you learned through observing, teaching, managing the homeroom class, and supervising students' school life. Also write about improvements to be made. 18. Evaluate the mentor teacher and your own student teaching activities. 19. Attend the student teaching evaluation meeting at your school and record the content of the meeting. 20. Respond to the student teaching program survey.

5 Evaluation

As mentioned earlier, the student teaching process is evaluated in many steps by the participants. In this section we provide the description of the evaluation process articulated in the student teaching portfolio (which can be accessed online at http://home.ewha.ac.kr/~ofe/).

There are three main evaluations in which the student teacher participates throughout the student teaching process. The first kind of evaluation is the one carried out in the post-teaching meetings. In this meeting, the student teacher, mentor teacher, other teachers, and other student teachers who observed the lesson evaluate the lesson and provide feedback. The post-teaching meeting of the lesson that the supervisor observes includes the supervisor. Second, as the end of student teaching comes, the school opens a final summative evaluation about the student teaching process. Here the teachers and administrators who participated in the planning and executing of student teaching and all the student teachers at the school discuss and evaluate the student teaching process. Third, after the student teaching is completed, the student teacher attends the evaluation meeting held at the institute and evaluates the student teaching program. This includes submitting the self-evaluation form.

There are three main evaluations that the mentor teacher participates in throughout the student teaching process. The first and the second are the same as the aforementioned evaluations that the student teachers do. The third is the evaluation of the student teacher that is submitted to the institute which provides important information for the final grade that the student teacher receives.

Finally, the supervising professor evaluates the student teachers and the student teaching process. Based on the evaluation collected from the mentor teacher, lesson plan, class observation, and student teaching portfolio, the supervising professor assigns a grade to each student teacher for the student teaching.

6   Research on the Perception of Student Teaching

Several studies have investigated how student teachers perceive their student teaching experience and researchers have suggested further improvements in developing student teaching programs[1,2,5,6,9].

In their study of secondary mathematics teacher candidates' perception of student teaching, Shim and Lee[9] found several results from the survey questionnaire they implemented with 119 teacher candidates who completed student teaching. Among the prospective teachers who responded in the survey following a seven point Likert scale: 89.9% agreed positively that the student teaching experience was a good opportunity to apply what their knowledge gained from the institutes to the actual field; 97.5% positively replied that the supervision from the mentor teacher was helpful; 96.6% positively replied that the observations from other math teachers and fellow student teachers were helpful in improving mathematical pedagogy. Also, in the questions asking about the change of perceptions through student teaching, 95.8% positively replied that through student teaching they felt less nervous about teaching students; 99.2% positively replied that teaching lessons helped improve pedagogy and the same percentage said they felt the joy and value of teaching students; and 95% positively replied that they were able to check their aptitude and reassure their passion towards teaching.

Jo[6] explored the educational values of student teaching through investigating the student teaching journals of 17 teacher candidates and by conducting additional group or individual interviews with the teacher candidates. He found that firstly, student teachers regarded their student teaching experience as a debut in the teaching field. Secondly, they felt a great joy and happiness when they had the opportunity to teach students despite being confronted with unpredicted and difficult situations. Thirdly, their ambiguous concepts of teacher, students, and schools were refined through their experience in schools to a more realistic view. Fourthly, the student teachers found student teaching an important experience of learning how to teach.

Chung and Chung[1] investigated how various student teaching experiences affected the development of key competencies required for a teacher in pre-service teachers. Through a survey conducted with 120 pre-service teachers, they found three main results. First prior to student teaching, the pre-service teachers expected to develop the various key competencies required for a teacher, especially, those competencies related to student supervision and teaching. Second, they identified four different types of mentoring provided by the mentor teachers and concluded that the types of mentoring affected the development and changes in the competencies of the pre-service teacher. Third, they found that the environment of student teaching which involved the type of the mentor teacher affected the development and changes in the competencies of the pre-service teacher. Further, Chung and Chung[1] suggested two elements for cooperating schools to consider a desirable student teaching environment. One is to select mentor teachers who command respect and can provide appropriate feedback. The other is to provide a maximum amount of time in the school teaching, as long as the school curriculum allows.

As one solution to provide more time in schools for prospective teachers, Ju[7] discussed the effects of a college student mentor program. Originally the mentoring program was established as a way to reduce private tutoring and to emphasize learning in schools. In this program, college students majoring in education or who are in other majors taking the alternative course (explained above in the requirements for student teaching), volunteer as mentors to elementary or secondary students in extracurricular and academic activities. They can also volunteer as a teacher aid for regular school classes. Although planned primarily for reducing private tutoring, this program is considered as another way to provide more time and opportunities for prospective teachers to gain experience in the actual field[7].

References

1. Chung, M., Chung, J. (2012). An analysis of the teaching practice effects on the competencies of pre-service teacher. *The Journal of Korean Teacher Education, 29 (4),* 63–83. [in Korean]
2. Lee, B. (2008). A preliminary study for the supervision of pre-service mathematics teachers. *Journal of the Korean School Mathematics Society, 11(1),* 1–18. [in Korean]
3. Lee, J., & Sohn, H. (2005). *The concept and practice of student teaching.* Goyang: Seohyunsa. [in Korean]
4. Na, S., Lee, M., Park, M., Han, H., Kim, I., Lim, H., & Hwang, Y. (2012). *A student teaching guide for pre-service teachers.* Paju: Gyoyukgwahaksa. [in Korean]
5. Jeong, H. (2013). A study on student teachers' perception of academic adviser in pre-service internship course. *The Journal of Research in Education, 26(2),* 77–114. [in Korean]
6. Jo, S. (2008). Educational values of teaching practicum as a part of teacher preparation course. *The Journal of Educational Administration, 26(2),* 77–114. [in Korean]

7. Ju, M. (2006). The effect of teaching experience in after-school learning programs: Implication for the development of mathematics teacher education program. *The Mathematical Education, 45(3)*, 295–313. [in Korean]

8. Seo, K., Joo, Y., Lee, J., Hyun, S., Lee, J., Shim, S., Kim, J. (2006). Developing a standards-based student teaching program. *The Journal of Korean Teacher Education, 23(3)*, 275–303. [in Korean]

9. Shim, S., Lee, K. (2013). Perceptions of pre-service mathematics teachers' teaching practicum and difficulties of mathematics instructions. *The Mathematical Education, 52(4)*, 517–529. [in Korean]

CHAPTER 4

MASTER TEACHER'S ROLE IN PROFESSIONAL DEVELOPMENT AND THE MENTOR TEACHER SYSTEM

Sanghwa Kim

Master Teacher, Yongin San-yang Elementary School
#2 Sanyang-ro, Giheung-gu, Yongin-si, Gyeonggi-do, 446-598, South Korea
E-mail: exit90@dreamwiz.com

In this chapter, I would like to introduce the newly established master teacher system in the Korean educational system. I will discuss the purpose of the master teacher system, and describe the role of the master teacher. I also provide a case of a master teacher specializing in mathematics in a public elementary school.

1 Master Teacher System in Korea

The Republic of Korea (South Korea) has a long tradition of placing high value on the teaching expertise. Over time this perspective shifted to a system in which bureaucratic tasks and administrative roles became highly value than instructional quality, and subsequently teachers who work hard outside of teaching were promoted to more prestigious administrative positions. This loss of focus on instruction as the original task of teachers gave rise to the master teacher system. The primary goals of the master teacher system are to enhance the status of teachers and teaching, and reduce the administration-oriented culture in which being an administrator is perceived as a promotion for a teacher. With master teacher system, teachers foster their expertise in teaching and curriculum by participating in high quality professional development and mentoring. Ultimately, it is expected that master teachers will play important roles in creating of a learning community where teachers share their ideas and learn together.

1.1 *Operation of Master Teacher System*

In 2011, the Korean Ministry of Education, Science and Technology proposed two directives for the master teacher system. First, the teaching profession is elevated in status through a rigorous process for selection of the master teacher. This includes an official recruitment system through competency evaluation, progressively reducing the number of classes that master teachers must operate, and providing government funding for research and related projects. Teachers who have Level 1 Teaching Certificates with more than 15 years of classroom experience can apply for the post of master teacher. These candidates must then go through a two stage screening process following national guidelines, administered annually at the regional level. The first stage of this selection process involves documentation review, and evaluation by the school administrator and peer teachers. The second stage includes evaluation of lesson plans and teaching in classroom to assess candidates' specialty in subjects. This stage also evaluates candidates' competency in the consulting arena using the classroom episode. Finally, an in-depth interview is used to evaluate candidates' overall knowledge of the educational system, their role as a master teacher, and their attitudes towards teaching.

Candidates selected through this extensive screening and selection process are appointed master teachers, provided 2-4 weeks of intensive qualification training during their school holiday, and finally, placed in schools who have requested a master teacher.

Second, as previously mentioned, the master teacher has reduced teaching hours so they can have more time to help their colleagues to enhance the colleagues' teaching expertise. The master teacher can provide support in the form of informal discussions, formal consultation and training, modeling and classroom demonstrations, and structured professional development activities. They are also involved in development of curriculum materials, and evaluation methods.

1.2 *Foreign Programs Similar to Korean Master Teacher System*

Similar programs to the master teacher system in Korea are shown in Table 4-1. To improve classroom teaching, the United Kingdom (UK) has Advanced Skills Teachers (AST) and Excellent Teachers (ET), Singapore has Master Teacher and Senior Teacher, USA has Support Provider and Mentor Teacher, China has First Level Teacher and France has professeur agrégé. These teachers are placed in schools, school district, or educational office. They are responsible for mentoring beginning teachers and prospective teachers, developing teaching methods, sharing exemplary teaching practices, creating teaching tools, providing in-service teacher training, and visiting classrooms to observe classes and discuss teaching practices (Ministry of Education, Science and Technology, 2011; Choi, 2011).

Table 4-1. Foreign Programs Similar to Korean Master Teacher System

Content	Name	Roles and Responsibilities
UK	Advanced Skills Teachers (AST)	◦ Advanced Skills Teacher -Supporting for classroom management and teaching -Mentoring beginning teachers and prospective teachers -Sharing exemplary teaching practices through on-site visits
	Excellent Teachers (ET)	◦ Excellent Teacher -Training beginning teachers -Mentoring teachers -Sharing best practices through demonstration classes -Supporting teachers in making a lesson plan, class preparation, and evaluation
Singapore	Master Teacher	◦ Master Teacher -Supporting teachers in the same school district in teaching methods -Supporting personality development of students -Fostering cooperative relationship with parents, community and educational experts
	Senior Teacher	◦ Senior teacher - mentor and role model for novice teachers in school -Delivering teaching expertise and curriculum -Role model as expertise in instruction and student learning -Training prospective teachers and beginning teachers
USA	Support Provider (LA Unified School District, Los Angeles, California) Mentor Teacher	◦ Mentoring new teachers

	(Kyrene School District, Tempe, Arizona)	
China	First Grade Teacher	∘ Enhancing the quality of elementary, middle school and high school teachers including demonstration class, observation class, academic lectures and class analysis ∘ Publishing more than one educational article biennially or presenting at academic conferences, and publishing textbooks
France	professeur agrégé	∘ Providing educational services mainly in high schools and teaching 15 hours per week ∘ Evaluating students' academic achievements, career counseling and parents' consulting

The roles of master teachers in Korea are similar to the combined roles of Excellent Teachers in UK, Master teacher and Senior teacher in Singapore, and First Level Teacher in China. Like Master Teachers in Singapore and First Level Teachers in China, master teachers in Korea provide support outside of their own school through offering demonstration classes, collegial consulting, professional development, participation in various academic societies and presentations at professional conferences. All of these with the goal of increasing the expertise of educators in the region. In their home schools, master teachers in Korea also mentor less-experienced teachers, providing classroom observations to assess and support in issues of curriculum and educational activities. These master teachers also offer open classes for other teachers to observe. These activities are similar to the roles of Excellent Teachers in the UK and Senior Teachers in Singapore.

2   In-School Activities of Master Teachers

2.1 *Math Classes of Master Teacher*

In Korea, all teachers are expected to open their classes to observation by visitors at least twice per year, typically between April and October. Schools make their own plans regarding the time and process of open classes. Because teachers often find it so stressful to open their classes and discuss their teaching with outside observers, they often resort to the default lesson plans provided by the school. Additionally, during these open observation periods, the master teacher may bring other teachers who teach the same subject matter to observe and discuss teaching strategies. Although classes taught by master teachers are always open for observation, teachers may find it hard to arrange time to do these visits. Therefore, to encourage teachers to observe classes, I have been inviting other teachers to observe my math classes by selecting a unit or theme, then assessing the concerns and needs of teachers. Additionally, I let teachers know that my class is always open and anybody can come and observe. The special open classes I have offered between 2011 and 2013 are shown in Table 4-2.

Table 4-1. Math Classes of Master Teacher

Year	Grade	Unit (Subject)	Contents	Comments
2011	4	Math class to enhance algebraic thinking	1. Line of large numbers	Presented diverse teaching methods and contents of lesson to help enhance algebraic thinking
			2. Comparing of multitude of large numbers	
			3. Multiplying 100, 1000, and 10000	
			4. Multiplying 100s, and 1000s	
			5. (Three digit number)×(Two digit number)	
			6. (Four digit number)×(Two digit number)	
			7. Multiplication of three numbers	
			8. Sum of the angle measures of a triangle	
			9. Operations using addition and subtraction together	
			10. Operations using multiplication and division together	
2012	5	Measurement of area of plane figures	1. Area of a rectangle and a square (reconstitution)	Presented a class focused on conceptual understandings and connections between mathematical concepts
			2~3. Area of a parallelogram	
			4~5. Area of a triangle	
			6~7. Area of a trapezoid	
			8. Area of diamond	
2013	6	Basic principles of dividing decimals	1. (Decimal number with one decimal place)÷(Decimal number with one decimal place)	Presented a class to help students with mathematical thinking and operation
			2.(Decimal number with two decimal places)÷(Decimal number with two decimal places)	
			3. Dividing two decimals with different decimal place	
			4. (Natural Number)÷(Decimal number)	
			5. Calculation of remainder in the division of decimals	
			6. Calculation of quotient by rounding off	
			7. Solution review in different mathematical situations (Reconstitution)	

The open classes of master teachers are designed not merely to show off how great a teacher is, but to illustrate how a teacher can interact effectively with students; eliciting participation and facilitating productive discussions. When the master teacher announces an open class session, they state their teaching objectives, as well as the content and focus of the class, so that teachers who observe the class can better understand what will happen. Some teachers attended multiple classes, while others visited once or twice. Teachers who have observed open classes of master teachers frequently reported that they learned that mathematical communication were possible even if master teachers were not homeroom teachers. These teachers also reported that they learned the importance of diagnosis of students' mathematical understanding prior to making their lesson plans.

2.2 *In-school Class Consulting*

Recently, the perspective of improving teaching has shifted from "top down" directives from a school district inspector to a more collaborative process whereby collegial consultation focuses on the problems posed by teachers cosidering classroom situation. Figure 4-1 below, illustrates the collaborative process I have used to bring all grade-level teachers together to participate in a discussion that I facilitate. These discussions are then combined with individual classroom observations and may include one-to-one consultation for those teachers who request it. In elementary schools, the mathematics master teacher in Korea is required to provide mentoring and consultative support in all subjects, even though the request may fall outside their mathematics teaching expertise. This can prove quite challenging at times.

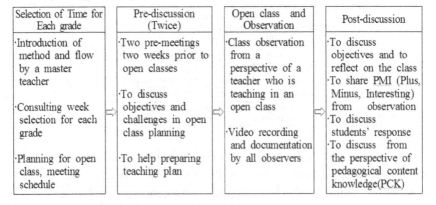

Selection of Time for Each grade	Pre-discussion (Twice)	Open class and Observation	Post-discussion
·Introduction of method and flow by a master teacher ·Consulting week selection for each grade ·Planning for open class, meeting schedule	·Two pre-meetings two weeks prior to open classes ·To discuss objectives and challenges in open class planning ·To help preparing teaching plan	·Class observation from a perspective of a teacher who is teaching in an open class ·Video recording and documentation by all observers	·To discuss objectives and to reflect on the class ·To share PMI (Plus, Minus, Interesting) from observation ·To discuss students' response ·To discuss from the perspective of pedagogical content knowledge(PCK)

Figure 4-1. The Collaborative Process In-school Class Consulting

2.3 *Mentoring Beginning Teachers*

As a master teacher, I have offered mentoring twice a month to improve the expertise of teachers with less than 5 years experience, and from time to time, provide more intensive private consultation on an individual basis. Each month, I gave guidance to teachers regarding mathematics knowledge to teach and mathematics classroom culture. Additionally, I facilitated monthly meetings where five novice teachers met together and openly discussed their own challenges. In this workshop, I tried to discuss a broad range of topics with which beginning teachers have difficulties, such as classroom management, student behavior, consultation with parents, and communication with students.

2.4 *Creating Teacher Community*

One of the major roles of a master teacher is to create a learning community with an atmosphere in which teachers can freely share their ideas and safely discuss their experiences related to classroom teaching. To accomplish this goal, I organized the "Class Review Club" comprised of 8 teachers who volunteered to critically review their own classroom teaching. In other words, the primary purpose was to learn to analyze our own teaching in retrospect and improve our teaching practices by developing a critical perspective. The group was also designed to obaseve other teachers' classes, write and share constructive critiques about the class, and apply the lessons to our own classes to improve our teaching. For the purpose, the members of the community analyzed and critiques classes according to the flowchart shown in Figure 4-2. Not all were matheamtics classes. To open the class, teachers decided the subject which they are interested in and want to discuss. At first, the master teacher opened her mathematics class and invited teachers.

To write an effective class critique, teachers as observers should clearly understand the intention of the teacher and the goal of the class. Then teachers wrote a narrative reflection on the class, considering the context and the classroom situation. The importance of such writing is not that teachers have chances to attend others matheamtics class, but that teachers have opportunites to reflect on their own teaching, and to share and discuss common issues in mathematics teaching.

Pre-discussion	· To explain teacher's intention of the class · To develop a lesson plan after discussing mathematics knowldege, student understanding, teaching methods, and problem posing · To ascertain key points of class observation and listen to a perspective of a teacher who is going to teach in an open class (may vary from class to class)

⇩

Open class and Observation	· Common observation criteria: Is the class consistent with the teacher's intention? Are students' participation, mathematical thinking, and communication valued in the class?(How is the mathematics classroom culture?) · Individual observation criteria: Teacher A & B– Do classroom tasks comply with objectives of the class? Teacher C & D– How are the reaction of students during the class? Teacher E & F–How are the reaction of low-achieving students? Teacher G & H- How are teacher's performance and how does a teacher pose questions? * Observation point may vary depending on objectives of a class

⇩

Writing Critique	· To write a factual critique as an observer focused on coherence and situation of the class · To write questions and what needs to be discussed from the open class in light of own experiences

⇩

Sharing critique	· To share and discuss based on each other's critiques · To record conversations during sharing process · Master teacher to facilitate discussion

⇩

Writing about critique sharing	· To document recorded sharing process in writing

⇩

Writing on self-reflection	· To reflect on the class through critique writing, sharing critiques · To apply the lessons to their own class to improve their teaching

Figure 4-2. The Process of Class Review Club's Activities

3 Extracurricular Activities of a Master Teacher

3.1 *Activities to Create a Meaningful Connection between Educational Policy and Mathematics Education*

In Korea, when the national government or local educational office establishes an educational policy, the goal is to implement the policy in the classroom as soon as possible. Currently, the Korean government emphasizes creativity and personality development in education, and the Kyunggi Provincial Office of Education, where I am belong in, focuses on learning oriented classes to enhance creativity.

When a new policy is launched inspectors, master teachers, and school administrators such as principals and vice principals participate in workshops to understand the educational policy fully. After the formal leaders have been trained, front-line staff are trained on the new policies. Unfortunately, although these policies may include changes in educational philosophy and curriculum, they rarely include specific suggestions for detailed changes in methodology or instructional practices in the classroom. The master teacher is responsible for making the critical connection between educational policy and classroom practice. This is done by critically evaluating the policy, applying their expertise to developing specific instructional strategies, field testing the policy in their own classroom, and then sharing them with their colleagues. Table 4-3 represents the steps I followed in applying educational policies of the Kyunggi Provincial Office of Education in my school.

Table 4-3. Activities to Create a Meaningful Connection between Educational Policy and Mathematics Education

Time	Major Activities	Details
Jun. 2012	Discussion on the learning oriented classes with special officials	· *Opened* a math class adopting the educational policy to special officials and had a discussion after revealing the intention of the class (Open Grade 5 math class regarding the area of rhombus)
Sep. 2012	Discussion on learning oriented classes with vice principals	· *Opened* a math class representing educational policy to special officials and had a discussion after revealing the intention of the class (Open 5 Grade math class regarding the shape of line symmetry))
Feb. ~ Mar. 2013	Development of criterion for class observation	· Participated in development of criterion for learning oriented class applying the intention of educational policy of Kyunggi Provincial Office of Education
Jun. ~ Aug. 2013	Development of program for educational policy training	· Developing program in the remote training to understand learning oriented classes
Apr. ~ Dec. 2013	Supporting corps activities	· Corps activities to support learning oriented class (Lecture and consulting)
Aug. 2013	Workshop (training teachers)	· Direction and examples of math classes adopting the policy · Discussion methods in math class

3.2 *Consulting Beyond My Own School*

In the Kyunggi Provincial Office of Education, regional educational support office provides an administrator who performs the role of consulting manager as shown in Figure 4-3, below. This manager serves as an additional support, and a liaison between the clients and consultants. Consultants are selected from a group of master teachers, who are certified in the relevant subject area, and administrators.

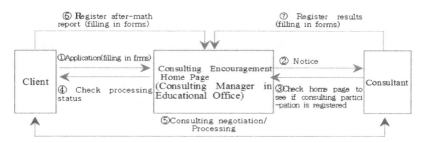

Figure 4-3. Examples of Instructional Consulting Model

Yongin Educational Support Office in Kyunggi-do provides a consultative support process as illustrated in Figure 4-4 below. There are three stages in the consultation process; Planning Stage, Implementation Stage & Evaluation Stage.

Consulting stage		Details
Planing	1. Preparation	1) Commission and reception
		2) Provisional diagnosis and consultant assignment
		3) Consulting agreement
	2. Diagnosis	4) To clarify commissioned task and check the purpose
		5) To collect materials and clarify necessary facts
		6) To analyze materials and set future direction
Implementation	3. To devise and select solutions	7) To devise and propose solutions
		8) To select solutions
	4. Performance	9) Execution and observation of the client
		10) Guidance and advice of the consultant
Evaluation	5. Closing	11) Consulting evaluation
		12) Final report preparation

Figure 4-4. A Consultative Support Process

The Planning Stage. Phase 1 (Preparation) of the Planning Stage is initiated when potential clients submit a request for support via the website of the Provincial Education Office (Step 1). The Education Office then designates an administrator to serve as the consulting manager. In Step 2, the manager assigns a consultant based on the provisional diagnosis, and schedules a meeting between the two parties. In Step 3 of the Preparation Phase, the manager facilitates a meeting between the client and their consultant with the goal of collaboratively developing a consultation plan. This plan is an agreement between the two parties outlining the purpose and goals of consultation, and clarifying professional roles during the consultation process. The Preparation Phase is followed by the Assessment or Diagnosis Phase in which the purpose is reviewed, assessment information is collected, and possible future directions are considered.

The Practice State. This stage of the process involves phases, planning and implementation. During Phase 3, assessment information is carefully considered, and potential solutions are proposed (Step 7). The client then selects the solution to be implemented (Step 8). Phase 4 involves the performance of the selected solution. In Step 9, the client executes the plan under the supervision of the consultant. Subsequent implementation problems are addressed by the personal guidance and advice of the consultant who has observed the class during implementation (Step 10).

In the Evaluation Stage, consulting is evaluated through satisfaction report of the client and self-evaluation of the consultant and evaluation report is made based on these. The report is then submitted to the consulting manager (Kim, 2013).

Consulting Cases Table 4 lists 28 consulting cases in which I have served as the consultant. All 28 cases involved issues in my specialization area of mathematics. Of the 28 cases, 19 teachers requested help on only one issue over three years. Three teachers required help on two issues. And 1 teacher required consultative support on three separate issues from 2011 through 2013. Sometimes, the Initial presenting problem was changed into a more concrete one during the diagnosis process as shown in Table 4-4. In general, most clients wanted to learn strategies for teaching math that were consistent with the policies of Kyunggi Provincial Educational Office. In order to better understand the implementation of these strategies, consultation often began by having the client observe me modeling them, followed by a debrief discussion of what they observed. In some cases, the issue was client mathematics content knowledge. In these situations, we would discuss the specific topic or unit, and devise a teaching plan based on teachers' mathematics content knowledge and students' mathematics understanding. After observing the teacher's mathematics class, we met to evaluate whether or not consultation met the client's goals (Kim, 2013).

Table 4-2. Consulting Practices in the Mathematics

No.	Time	Sex	Experience (Years)	Grade	Theme	Tasks inferred from diagnosis
1	2011	F	3	5	Circumference and area of a circle	Consulting on the predesigned teaching plans, comparison between learning oriented class perspective and my own class, and practical implementation method to implement teaching according to level
2	2011	F	4	5	Cuboids and cube hexahedrons	Math class in accordance with the intention of learning oriented class, teaching method according to level of students, general things of class composition
3	2011	F	4	5	Area of parallelograms	Methods to make classes with active interactions with students, regaining confidence in math classes
4	2011	F	30	5	Area of rhombus	Check whether the designed teaching plan matches with learning oriented teaching policy, effective motivation and cooperative learning method
5	2011	F	7	6	Understanding and Interpretation of circle graph	Check if constant activities to enhance mathematical communication abilities are meaningful (my math class culture)
6	2011	M	5	6	Proportional ratio	Check if the introduction is helpful to understand a concept, configuration of contents to make them understand the concept of proportional ratio, and expect wrong concepts and students' responses
7	2011	M	7	4	parallelogram	Consider connectivity between triangles and rectangles, how to configure contents to make students find thee concept, learning contents evaluation method
8	2011	F	29	2	Evenly divided figures	Class flow for clear understanding of even division, devise teaching materials and reconfigure teaching contents
9	2011	F	15	5	Fraction division	Review of the principle and various methods to convert the division of fractions into multiplication, cooperative learning method and concrete introduction of the teacher
10	2012	F	4	4	Prime numbers	Configure contents to allow students to learn concepts of prime numbers through their activities
11	2012	M	17	6	Division of decimal numbers in real life	STEAM integrated educational course and storytelling application (reconfiguration) in theme explorative math classes
12	2012	F	25	5	Multiplication of three factions	Math classes that can be called as learning oriented classes and cooperative learning with continuous interest and meaning

13	2012	F	15	1	Comparison of volume	Review the goal and level of Grade 1, activities and learning method, proper teaching materials, and evaluation in the class
14	2012	F	26	2	Chart	Class flow applying ARCS motivation strategy, class contents considering creativity enhancement, dealing with real life to enhance the effects of teaching materials
15	2012	F	26	4	Comparison of fractions with the same denominator	Activities and materials that students can understand the comparison of fractions with the same denominator
16	2012	M	17	5	Finding a rule through real application	Suitability of real life learning materials, activities and introduction to review their own solutions and others' solutions with critical thinking
17	2012	F	18	1	Length comparison	Length comparison activities suitable for Grade 1 students, introduction and teaching method suitable for Grade 1 students, selection of teaching materials, general class flow
18	2012	F	30	2	Make problem according to formula	Effective method to make problems and share them one another, whether to match with the educational policy
19	2012	F	19	4	Rhombus	Class flow to get to know rhombus, introduction of the teacher to induce students' communication
20	2013	M	18	4	Mixed calculation of three numbers	To provide questions based on the materials familiar with students to motivate and review math course for the class
21	2013	F	2	4	Mixed calculation of three numbers	To check if it is possible to give math class through discussion among students, and how to make it possible
22	2013	M	11	4	Circle and the ratio of the circumference of a circle to its diameter	To review contents knowledge on circle and the ratio of the circumference of a circle to its diameter, methods to make students experience, feel, and realize
23	2013	F	19	1	Weight comparison	Expectation and countermeasure of diverse responses that students show in the class
24	2013	F	31	2	Classification according to criteria	Review textbooks, and class flow for meaningful activities and to make them arrive at the goal of the class
25	2013	F	20	4	Relationship between a mixed number and an improper fraction	How to reconfigure class contents and amount, class flow making them know the relationship between a mixed number and an improper fraction
26	2013	F	3	6	Division of decimal numbers	Contents knowledge on the principle of division of decimal numbers, effective class operation method in the area of number and operation
27	2013	F	3	5	Division of decimal numbers	To learn the principle of division of decimal numbers, how to provide various problems to children who understand the principal but feel difficulties in calculation
28	2013	F	3	5	Graph and description	Application class to apply graphs that have learned, and cooperative learning and self-driven learning

3.3 *Consultation for Evaluation of Math Achievement*

Recently, Korea has focused on improving the system for evaluating student academic achievement. The emphasis has shifted from single answers to the evaluation of high-level thinking capabilities that better prepare children to be a part of the future of Korea. This is reflected in the increased use of open-ended and essay type questions to better reveal the thinking process of the student, as well as the resulting answer or conclusion. Additionally, the focus has shifted from "snapshot" summative assessment of achievement to the continuous formative assessment that informs instructional practice.

Due to the changing trend in assessment policy, many schools are moving towards more open-ended, in-depth tests combined with the informal assessment and professional judgment of teachers. As a result of these changes, many veteran teachers do not have the skills or knowledge to meet the assessment standards outlined in new policies. This leads to schools requesting consultation support for professional development on the topic of more effective math assessment aligned with government policy. Such consultation typically involves reviewing current teacher-developed assessments for meeting their stated evaluation goal and validity, and providing feedback and guidance.

4   Research-to-Practice and the Master Teacher

Effective math education is the result of skillful application in the classroom of the evidence-based ideas of researchers at universities and professional organizations. The translation of research-to-practice is critical in math instruction, as it is in many professional endeavors. This begins with ensuring that professional development for teachers of mathematics includes high quality instruction in both theory and practice of their subject matter. Nonetheless, even the most elegant theories and best instructional practices will fail to improve student outcomes and thus lose value, if teachers cannot perform them in the typical classroom.

The role of the master teacher in the research-to-practice process involves staying current on the latest research through reading professional journals and attending lectures. They then relay to front-line teachers the information they believe to be most important and effective. Additionally, master teachers deliver lectures and publish articles or books to share their accumulated professional experience.

4.1 *Activities in Math Education Related Societies*

In Korea, academic societies are organizations primarily comprised of university professors focused on the research and theory in their field. In contrast, professional associations are less formal, regional "study groups" primarily comprised of practitioners focused on following current policy and applying best practices in their school or classroom. As a master teacher, I tend to identify recent research trend in math education by participating in various academic societies related to math education, and reading journal articles dealing with elementary school education. Additionally, I participate in research projects focused on my experiences as a classroom teacher as well as those of a master teacher in the consultant role. These articles are then submitted to various academic societies for presentation and publication. Recent topics of research have included, an analysis of the characteristics of the consulting process in elementary math education, and teachers' perceptions and use of manipulatives provided with a mathematics curriculum.

Research recently presented at an international conference include an analysis of teachers' conception of the purpose of teaching mathematics reported in PME 35 insert space (The International Group for the Psychology of Mathematics Education) in 2011, and Korean elementary teachers' orientations and use of manipulative materials in mathematics textbooks reported in ICME-12 insert space (International Congress on Mathematical Education.

Most of members of math education related societies are university professors majoring in math education. Recently, there has been increasing demand for teachers in elementary, middle and high schools to participate in the academic society to make a balance between theory and practice. Consequently, teachers have become board members in the Korean Society of Mathematical Education and the Korean Society of Educational Studies in Mathematics. I am currently a board member of the Korean Society of Educational Studies in Mathematics. In the future, it will be necessary for teachers majoring in math education to participate in academic society activities so that theory and practice can be developed in balance, and the interest of teachers be represented and promoted.

4.2 *Activities in Math Education Related Research Associations*

As stated previously, Korean research associations are more regional and informal groups of front-line teaching professionals who get together to develop professional networks, share successes and challenges, exchange teaching materials and strategies, stimulate thoughtful introspection of beliefs and practices, and offer seminars and workshops. I have been a member of Kyungin Elementary Math Education Research Association since I started teaching. Membership of this group includes professors in Kyungin Teachers' College and teachers in Kyunggi Province and Incheon Metro City. The most important goal of this association is to help teachers build their students' mathematical thinking and communication skills. Subgroups of the Association meet monthly to discuss actual class teaching cases. In summer, there is an intensive seminar in which all branch members meet together, observe an open class, discuss class teaching, and taking a special lecture offered by a professor. The Association also organizes a Korea-Japan joint seminar with Tokyo Elementary Math Education Research Association in Japan every 18 months. In this joint seminar, Japanese teachers teach Korean students and Korean teachers teach Japanese students. With this special experience, we try to learn about math education in Korea and Japan through discussion after the exchange class.

Additionally, through the Korea Elementary School Master Teachers' Research Association and the Kyunggi Province Master Teachers' Research Association, master teachers in the Mathematics Division share resources via an Internet blog with the goal of enhancing their consulting activities. The Korea E.S.M.T.R.A plans to develop regular meetings by content area in the future, as well as offering professional seminars. In the Kyunggi Province Master Teachers' Research Association, we informally share the challenges, critiques and insights of classroom teaching seminars and job training.

5   Future Direction of the Master Teacher System in Korea

Classroom teachers of mathematics often feel a gap between theory and practice, between the theory of educational policy and the reality of the classroom, and between requests by administrators and the demands of the curriculum and the needs of students. Master teachers must truly understand the challenges of the classroom teacher and advocate for their needs, while simultaneously understanding the role of good theory and guiding policy, and advocating for the application of both in classroom instruction.

As a master teacher in Korea, I am keenly aware of the role I play in the research-to-practice process. As a teacher, I experience and understand the many challenges of managing a classroom and providing high quality instruction. As a master teacher, it is my responsibility to communicate these challenges in a way that researchers, policymakers and administrators can play a role in helping to solve them, while also communicating to frontline teachers how best to utilize the existing research and implement educational policy.

Additionally, master teachers can support their fellow teachers by offering a model classroom for observation, and encouraging them to improve their teaching practices by participating in the valuable learning activities of professional associations. The master teacher can model the process of continuous learning, thereby inspiring others to reflect and grow in their own practices.

In Korea, curriculum has changed many times, forcing teachers to constantly learn and adapt to new textbooks and content. This is particularly difficult for elementary school teachers who teach several subject areas. As a result, master teachers can help teachers to understand such changing curriculum and textbooks, including the intention of new curriculum and help them to offer meaningful math classes. Additionally, master teachers can help classroom teachers identify difficulties in actual application and to improve their knowledge and skills. Collegial consultation and the master teacher relationship can help cultivate a learning culture in the school that contributes not only to the effectiveness of all teachers, but increases the joy and confidence in the teaching experience. To be effective in these varied roles, master teachers must be equipped with competencies in both theory and practical skills. Their success also depends on other educators understanding the value of their role in improving student outcomes. The master teacher system is relatively new in Korean education. In order to take this valuable system to scale and ensure sustainability, government officers and educational leaders must create the necessary supporting structures. We must learn from the best practices found in similar systems in foreign countries, provide critical resources, create an awareness plan for educators, and develop essential supporting policies.

The master teacher system can be a critical element of the continuous learning process that is necessary to prepare all students for a brighter future.

Acknowledgement

I would like to thank Dr Dale R. Myers who contributed to the revision of this chapter and translation into English.

References

1. Choi, J. (2011). The Master Teacher System of France. In J. Hyun, J. Hwang, & M. Choi (Eds.), *The Education Policy*, 219(pp. 30-33). Forum Korean Educational Development Institute. [in Korean]
2. Kim, S. (2013). Analysis of Characteristics of Commissioned Task of Elementary School Mathematics Teaching Consulting. *Korean Association for Learner-centered Curriculum and Instruction, 13*(5), 613-635. [in Korean]
3. Ministry of Education, Science and Technology (2011). Foreign Cases Associated with Master Teacher System. *Press release(February 25, 2011)*. [in Korean]

CHAPTER 5

GIFTED EDUCATION IN SOUTH KOREA

Mangoo Park
Mathematics Education
Seoul National University of Education
1650Seocho-Dong, Seocho-Gu, Seoul, Korea
E-mail: mpark29@snue.ac.kr

This chapter addresses the gifted education in South Korea. One of the main characteristics of gifted education in Korea is that the Korean government has a systematic and centralized system. During the last decade, Korea has rapidly developed comprehensive gifted education programs. However, there are still many challenges to increase the size and scope and enhance the quality of programs in gifted education. As an overview of gifted education, this chapter consists of the goal and policy, history and present status, identification and selection process, institutions and programs, teacher education and support systems of gifted education in Korea.

1   Introduction

Although the systems of gifted education vary across every country, it is important for every nation to raise gifted human resources for strengthening the international competitiveness. In particular, the demand of creative gifted humans is increased in this era of rapidly developing science and technology. As a nation that lacks natural resources, Korea should mostly depend on creative human resources to lead the advanced development of science and technology. Thus, along with Korean parents' enthusiasm, the Korean government has increasingly invested in gifted education. However, the history of formal gifted education in Korea has not been long compared with that of other nations[6]. The Korean government regulates the system of gifted education, which has strengths and weakness.

This chapter overviews gifted education in South Korea focused on the education for mathematically gifted students. To explain the mathematically gifted education system in South Korea, I provide the goal and policy of gifted education, an overview of the history and present status of gifted education, identification and selection process, institutions and programs, teacher education for gifted students, and the administrative support system.

2   Goal and Policy of Gifted Education

According to the *Gifted Education Promotion law* of Korea, the definition of a gifted and talented person is "an individual who requires special education to develop innate potential with an outstanding talent." Accordingly, the goals of gifted education are to help gifted and talented persons to develop their innate potential by suitable education. With the education, they can contribute to the development of society and nation as well as promote self-actualization.

According to the Article 31.1 of the Korean constitution and Article 3 and Article 19 of the Fundamentals of Education Act, all people of a nation shall have right to be educated according to their abilities and aptitudes to promote self-actualization and contribute to development of society and nation. Under the Constitution, most of the gifted education programs are also aimed at helping gifted and talented students develop creativity, leadership, teamwork, and self-directed learning ability.

Figure 5-1. Three Main Goals of Gifted Education

Having a vision of optimizing of upbringing of creative humans, the Ministry of Education (MOE) set up three main goals of gifted education: increment of gifted education opportunity, improvement of gifted education quality, and enrichment of gifted education effectiveness (see Figure 5-1).

Recently, as shown Table 5-1, MOE announced five areas and seventeen initiatives in gifted education as the 3rd Promotion Plan (2013 ~ 2017).

Table 5-1. Areas and 17 Initiatives in Gifted Education

5 Areas	17 Initiatives
Increase gifted education opportunities for nourishing dream and aptitude	Promotion of diversification and association of gifted education
	Elimination of blind spots of benefits in gifted education
	Improving the validity of selection methods
Substantiality of operating gifted education institutions	Reinforcement of relationship among gifted institutions at high school and university
	Diversification of gifted education institutions
	Improvement of supporting environment of gifted education institutions
Providing consumer-oriented gifted education courses	Providing consumer-fit gifted program
	Developing content for creativity and convergence
	Promoting content and operating system of gifted education program
	Reinforcement of quality management system of gifted education program
Reinforcement of educating able teachers and supporting system	Diversification of teaching resources for gifted education
	Professional development and capacity building of teachers
	Promoting supporting environment of gifted education
Building stable development basement	Efficiency of administrative supporting system
	Reinforcement of information and exchange support
	Increment and diversification of financial support
	Internationalization of gifted education

The main goals of this plan are summarized as 1) increment of gifted education opportunities for nourishing dream and aptitude, 2) improvement of quality of gifted education, and 3) reinforcement of effectiveness of gifted education. To bring up creative human resources by optimizing gifted education, this promotion plan will guide the directions of gifted education in Korea for the next few years.

3   History and Present Status of Gifted Education

It has not been that long since formal gifted education in Korea has started. With establishment of a science high school in 1983, Korean government drew much attention to gifted education. The necessity of strengthening gifted education has constantly been suggested since the submission of the Educational Report for the President by the Education Reform Committee in 1995. It was suggested that a gifted education center should be set up and operated in universities and research institutions. Since 1997, the Ministry of Science and Technology has supported gifted education centers attached to universities. An actual initiation of gifted education for students started with the enactment of gifted education promotion law (2000) and the enforcement of the law (2002).

MOE set up two comprehensive plans for the Promotion of Gifted Education: the first period from 2003 to 2007 and the second 2008-2012. During the first period, Korea Science Academy was established. During the second period, aside from the gifted education centers and schools, MOE expanded gifted education classes to general schools, with a goal of increasing of the number of students who receive gifted education up to 1% of the entire student population. The government will increase the number of gifted students up to 2% of the entire student population by the end of 2012 and 5~10% by the end of the 3rd Promotion Plan (2017).

Table 5-2. Status of Gifted Education Institutes[3]

Type		2008	2009	2010	2011	2012
Gifted Classroom		581	967	1506	2238	2520
Gifted Institution	Office of Education	226	251	258	354	261
	Attached to university	39	41	55	61	63
Gifted High School		1	2	3	4	4[a]

As shown in Table 5-2, the number of gifted education institutions has increased. Accordingly, the number of students who attend gifted education programs has increased. In 2012, 118,377 students attended gifted education programs during weekends as well as summer and winter vacations at gifted education institutions or gifted classes.

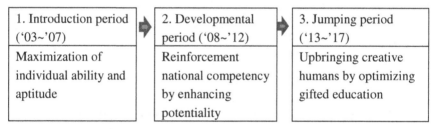

1. Introduction period ('03~'07)	2. Developmental period ('08~'12)	3. Jumping period ('13~'17)
Maximization of individual ability and aptitude	Reinforcement national competency by enhancing potentiality	Upbringing creative humans by optimizing gifted education

Figure 5-2. The Master Plan of Gifted Education

MOE set up three phase of Master Plan (see Figure 5-2): 1) Introduction period ('03~'07) focused on maximizing individual ability and aptitude, which emphasizes individual potential development, 2) Developmental period ('08~'12) focused on the reinforcement national competency by enhancing potential, which includes expanding opportunities of gifted education to more students and training opportunities for teachers and enhancing the quality of gifted education, and 3) Jumping period ('13~'17) focuses on upbringing creative humans by optimizing gifted education. With the development of various programs and the enhancement of gifted education, the government will

---

[a] Currently there are 6 Gifted High Schools in Korea[2].

support to bring up most creative humans in mathematics and science field and raise national competency.

4 Identification and Selection Process

The identification and selection process of gifted education in Korea has changed to pursue a better identification process. Before 2007, the identification process consisted of school principal recommendation, logical thinking tests, creative problem-solving tests in the field of mathematics, science, and information science and personal interview.

According to the '2nd Master Plan for promotion of gifted education' (2008~2012), the direction of identification was modified from paper-based tests to various methods. The process is composed of 4 steps in general.

- •Step 1: Teacher recommendation
- •Step 2: Giftedness test
- •Step 3: Specific academic aptitude test
- •Step 4: Personal interview

The identification process included more realistic approaches such as recommendations by teacher observations with various observation records. In addition to teacher's recommendation, the process also includes gifted tests that enhance creativity and intelligent motivation, specific academic aptitude tests in mathematics, science, information science, arts and so forth, and finally in-depth personal interview with or without mathematics and science camp.

The current trend is that steps above are unified together for selecting gifted students for the placement of the gifted classes or institutions. Figure 5-3 shows the identification and selection process of gifted students who should qualify for the gifted education at the centers attached to universities.

Figure 5-3. The Identification and Selection Process of Gifted Students

The first step is recommendation by an individual nominator such as the homeroom teacher, subject teachers, gifted education teachers or related professionals. This method has strengths in that school teachers have realistic observational evidence over a long period of time on the mathematical giftedness of the students.

The second step is recommendation by school recommendation committee. The school committee reviews teacher recommendation letters and various documents that show students' giftedness and recommend the student to selection review committee.

The final step is review by the committee organized by gifted education institutions. The committee prepares for their own identification criteria that include school recommendations, observational documents, personal interviews, and creative ability test results on giftedness.

5 Institutions and Programs for Gifted and Talented Students

There are three kinds of formal institutions for gifted and talented education: Gifted classes operated by each school, gifted education centers operated by offices of education, and gifted education centers operated by universities. As shown Table 5-3, gifted students who attended gifted classes from grade 4 to 6 at primary level and grade 1 to 3 at secondary level can transfer to gifted centers or institutions at universities. Most students who attended gifted programs at middle school level usually pursue science academy or science high schools that have high level of mathematics and science programs. Also, those schools are attractive for most students because the students who graduate gifted high schools or science high schools have advantages entering colleges or universities.

Table 5-3. Institutions of Gifted Education

Institution	Program	Operation	Fund
Office of Education	Gifted Class	Gifted class in each school	Regional community or individual support
		Class composed of students near the class	Gifted education center or individual support
	Gifted Education Center	Office of education	Operated by office of education
		Science high school, Science museum, etc.	Science high school (Science Academy), Science museum funded by government
University	Gifted Education Center	Ministry of Education	27 universities supported by Korea Foundation for the Advancement of Science & Creativity
		Office of Education	Local office of education

The characteristics of gifted classes and institutions are as follows.

Gifted Class: It is operated by elementary, middle and high schools in the form of extra-curricular activities, discretionary activities and after-school activities during week days, weekends, and vacations. They have their own school unit programs or class programs among schools that are joined by several neighboring schools.

Gifted Education Center: This center is usually operated by universities, government-funded research institutions, and public-service corporations. In addition to the regular classes at the normal school, students mainly attend these gifted centers on weekends or during summer and winter vacations. In general, students and their parents favor to attend these centers, in which they can learn higher level of mathematics compared to those of gifted classes. Each center is located near to residence of students so that they can go to the centers in a short period of time.

Gifted High School (Science Academy): It is operated by a full-time system in a high school level to bring up the highest potential abilities. The government funds this school, and graduates from these schools have advantages in entering colleges and universities.

According to Laws of Promotion of the Gifted Education, Korea has a systematic and centralized gifted education system. Figure 5-4 shows that students can attend either gifted classes or education centers at elementary and middle school level, schools for gifted students at high school level. After graduation of high school, the gifted students can receive special advance education. To provide gifted education, each school or center should have permission from MOE or city or province office of education.

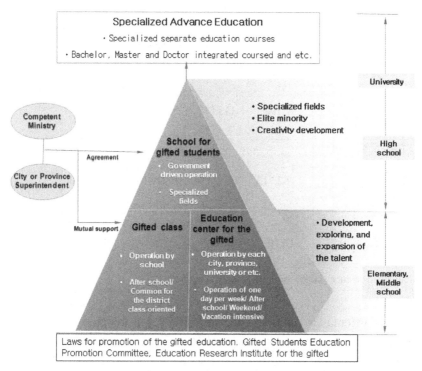

Figure 5-4. Gifted Education System by Institutions and School Levels

The number of gifted students are increased by years as shown in Table 5-4. According to the 3rd Master Plan, the number of students who will attend gifted education classes or institutions will be increased up to 5% of the total primary, middle, and high school students.

Table 5-4. Number of Gifted Students by Years

years	2005	2006	2007	2008	2009
Number of gifted students	31,100	39,011	46,006	58,953	73,865
Total number of students from elementary to high schools	7,757,900	7,724,840	7,757,023	7,617,047	7,387047
Ratio	0.40%	0.50%	0.59%	0.77%	1.00%

years	2010	2011	2012	2013	
Number of gifted students	92,198	111,818	118,377	121,433	
Total number of students from elementary to high schools	7,262,715	7,012,196	6,721,176	6,481,492	
Ratio	1.27%	1.59%	1.76%	1.87%	

*Students in science schools are included from 2008.

The curricula or programs for the gifted and talented students are basically made by institutions and schools. The curricula or programs at science centers for the gifted and talented students mostly consist of mathematics, science, and invention or information science. Table 5-5 shows the percentage of number of subject area programs that are included in the every type of gifted and talented programs.

Table 5-5. Number of Programs of Gifted Education[4]

Science	50.6%	Leadership	6.6%
Mathematics	45.8%	Creativity	6.3%
Invention	12.9%	Art	5.1%
Information Science	10.8%	Thinking Ability	4.8%
Integrated	8.1%	Music	3.6%
Foreign Language	7.8%	Others	3.3%
Language Arts	7.2%	Motivation	1.8%
Liberal Arts	6.6%	Physical Education	0.9%

Currently, instructors in each institution and center make their own teaching materials even though they have guidelines to follow when they help gifted students learn mathematics. However, the government recognizes the necessity of curriculum standards and initiates to develop a national standard program for gifted education and disseminate them to all the institutions and centers by 2017.

6 Teacher Education and Support System for Gifted and Talented Students

In South Korea, nearly all aspects of teacher education programs and the certification process are governed by the Office of Education or Ministry of Education. Government, offices of education, and institutions related to gifted education offer various types of professional development.

The teacher education programs consist of 3 steps as shown in Figure 5-6. The steps consist of basic course, enrichment course, and professional course. The basic course is operated by on site or cyber education for less than 60 hours for mostly pre-service or novice teachers. The enrichment course consists of program development and field application as team projects for more than 120 hours for experienced teachers in the field of gifted education, which is most important training among programs approved by MOE for advance certification. As the last step of teacher education program, professional course consists of thematic education and lesson and assessment consulting for more than 90 hours for experienced teachers and the professional coordinators of institution operation in the field of gifted education.

Table 5-6. Teacher Education Steps for Gifted Education

Step	Basic Course	Enrichment Course	Professional Course
Central and operation	• Basic course • Cyber or on-site education	• Strengthen program development and field application ability • Team project	• Thematic education (teaching method, institute operation, evaluation, etc.) • Lesson and assessment consulting
Recommended hour	less than 60 hours	more than 120 hours	more than 90 hours per topic
Career and type	• pre or novice teachers • Gifted teachers or administrators	• 5 year of teaching experience of gifted students • teacher and instructor of teaching experience of gifted students	• 5 year of teaching experience of gifted students • professional coordinator of institution operation

Provided mostly by city or province offices of education, universities, or gifted education research institutes, the basic mandatory teacher training courses include understanding gifted and talented children and education, identifying giftedness for selecting students, and so forth.

Provided by city or province offices of education, or gifted education research institutes, the enrichment or intensive training courses include in-depth understanding of a talent and gifted education, method for using inspection tools, mathematical contents, various teaching and evaluating methods and development of programs for the mathematically gifted.

Offered by gifted education research institutes, the professional training courses include identifying gifted education and policy directions, selection of gifted students, training teachers and understanding of institutional evaluation.

Offered by MOE, city or province offices of education, or gifted education research institutes, the manager training course includes reviewing gifted education and policy directions, selection of gifted students, training teachers and understanding of institutional evaluation.

For effective education for the gifted students, three main administrative support systems cooperate each other. MOE establishes a basic policy and adjusts the policy, supports and funds institutions and centers of gifted education. City and provincial office of education train teachers, operate gifted education institutions, and educate professionals and parents of students. Two main research institutions such as Korea Foundation for the Advancement of Science and Creativity (www.kofac.or.kr/) and Gifted Education Research Institute (http://gifted.kedi.re.kr) develop gifted education programs and human resource projects, do basic research for gifted education, develop screening tools, materials, and educate teachers.

7   Closing Remarks

South Korea has a very centralized system for educating mathematically gifted students. Over the last decade, it has progressed in developing a comprehensive national gifted education programs. However, as a nation that has limited natural resources, creative human resources are more demanding in the competitive economies. Thus, gifted education should be focused on bringing up creative potential of students to effectively cope with international competition.

However, there are still many challenges to overcome to increase the size and scope and the quality of programs of gifted education. To overcome these challenges, as Choi & Do[1] and Sheffield[5] mentioned, the Korean government needs to transform the current programs and educating methods for gifted students to foster flexible and creative ideas and convergent thinking using STEAM (Science, Technology, Engineering, Arts, and Mathematics) programs. Also, the gifted education programs should be expanded to informatics, arts, physical education, creative writing, humanities, and social sciences as well as mathematics and science.

In addition to national aspects of gifted education, gifted education should support students' various dreams and aptitudes by finding individual potentiality, which ultimately leads to individual happiness. Also, various gifted education programs for alienated students such as students who have multicultural backgrounds or low socio-economic status should be prepared to improve their socio-economic status. In particular, character education and career exploration are included in one of the purposes of education programs for the gifted. Teachers or instructors should have concrete realization plans for these aspects.

For an effective system of gifted education, in addition to develop an effective identification process, it is important to make sure that all policy departments are linked and cooperate to employ a more systematic approach to recruiting, training, and allocating specialist teaching staff and to improve the quality of programs for each gifted student. This can be achieved by means of a better curriculum, more efficient teaching, and consultancy support to teachers.

References

1. Choi, Y. & Do, J. (2008). *Research on the characteristics of mathematically gifted students in Korea.* 11th International Congress on Mathematical Education, 6 July – 13 July, 2008, Monterrey, Mexico.
2. Kim, J. (2014). North asian special schools: Korea. In B. R. Vogeli (Ed.), *An international panorama of secondary schools for the mathematically talented programs and practices.* Singapore: World Scientific Press.

3. Ministry of Education [MOE] (2013). *The 3rd comprehensive promotion plan for gifted education (2013~2017)*. The Ministry of Education Report. [in Korea].

4. National Research Center for Gifted and Talented Education [NRCGTE] (2012). *GIfted education in Korea*. Retrieved on January 10, 2014 at http://gifted.kedi.re.kr/khome/gifted/gedEng/select.do.

5. Sheffield, L. (2012). *Mathematically gifted, talented, or promising: What difference does it make?*12th International Congress on Mathematical Education, 8 July – 15 July, 2012, COEX, Seoul, Korea.

6. Vogeli, B. R. (2014). *An international panorama of secondary schools for the mathematically talented programs and practices.* Singapore: World Scientific Press.

CHAPTER 6

QUALIFICATIONS OF TEACHERS IN SPECIALIZED SECONDARY STEM SCHOOLS IN KOREA

Kyong Mi Choi

Mathematics Education, Department of Teaching and Learning,
College of Education, the University of Iowa, Iowa City, IA 52242.
E-mail: kyongmi-choi@uiowa.edu

Jihyun Hwang

Mathematics Education, Department of Teaching and Learning,
College of Education, the University of Iowa, Iowa City, IA 52242.
E-mail: jihyun-hwang@uiowa.edu

The purpose of this chapter is to provide information about teachers in secondary specialized STEM schools – Science High Schools (SHSs) and Science Academies (SAs) in Korea. As a result of data analysis on teacher qualification and professional development programs in SHSs and SAs, the two types of specialized STEM schools are based on different legal regulations for teacher qualification. SAs and SHSs offer various professional development programs in order to improve quality of teaching for gifted students in STEM. However, schools ought to pay special attention to retain high quality teachers as well as clarify expectations for their teacher qualifications.

1 Introduction

Learners construct mathematics knowledge through engaging in many opportunities to learn such as mathematics instruction, curriculum, teacher expectation and belief for students etc[10]. Particularly, teachers have a significant responsibility to provide students proper opportunities to learn in school settings[29]. Research on teaching and teachers is important to offer opportunities to understand how to improve students' learning experiences and achievement. In gifted education, teacher's roles on students are critical due to teachers' involvement in identification of individual giftedness, providing learning experiences beyond regular curriculum, making recommendations, and guiding students to advanced content in addition to what is generally offered[3,25]. Teachers of the gifted and talented studnets have more roles to play as Chambelin and Chambelin[1] and Rogers[24] emphasized required lessons on educational practice in gifted education. For example, studies suggest various forms of subject-based and grade-based acceleration for gifted students and instructions for the talented students aligned with their achievement and interests because of the needs for differentiated and student-centered instructions, which can meet their educational needs as well as opportunity to socialize and learn with like-ability peers.

Although there are some research studies on what are generally required to teach the gifted students, there is a lack of studies on qualifications that teachers of the gifted should have as a preparation to teach gifted learners, especially students in specialized STEM (Science, Technology, Engineering, and Mathematics) schools. Teachers' roles and qualifications to teach gifted students in STEM areas are important to examine in many aspects: to understand who can become teachers of specialized STEM schools, to prepare pre-service teachers who want to teach at STEM schools, and to provide teachers continuing professional development.

It is well documented that Korean students' high performance in mathematics assessment in international perspectives such as TIMSS and PISA[9,21] and possible reasons of Korean students' high performance[5,15]. Also, Korean teams at International Mathematical Olympiad (IMO) consistently showed success in recent years 3rd through 5th places through 2005 to 2008, respectively[2] and the first in 2012. It is important to understand what contributes to their success in mathematics and science achievement as well as that of the mathematically gifted. So far, there are insufficient studies on mathematics learning and teaching the gifted in Korea and their teachers have been disseminated.

To understand STEM school teacher's roles and qualifications, it would be a good starting point to examine current teachers' qualifications and expected qualifications for the teachers in STEM schools. The purpose of this article is to provide information about teachers in two types of Korean secondary specialized STEM schools —

Science High Schools (SHSs) and Science Academies (SAs). To be specific, we attempt to answer the following two questions to discover what is required to teachers in specialized STEM schools for high-quality gifted education: (1) What qualifications secondary STEM schools in Korea want for their teachers and (2) What qualifications the current teachers in secondary STEM schools have and what efforts the schools make to enhance quality of teachers.

1.1 *STEM Schools*

Thomas and Williams[31] studied a status of specialized STEM schools in the United States and found that they are reactive to social, economic, and political problems like workforce crisis and international economic competition. However, they noted that current specialized STEM schools need to provide following roles: (1) supporting the interests of students whose needs are not addressed and met in a traditional secondary school setting; (2) providing opportunities to enhance learning and teaching STEM subjects; and (3) developing the skills of students in areas that will support national economic and technological powers. Benefits of specialized science schools for students were well documented. One of such studies is conducted by Stephens[30] through gathering comprehensive information of 11 state-supported, residential mathematics and science high schools in the United States in regards to students, faculty, curriculums and extracurricular activities. The results indicated that students in the specialized mathematics and science high schools experienced interaction with students who have similar abilities and meaningful research as well as college life and college-level study earlier. Stephens[30] concluded, "The enriching, challenging, and accelerated experiences that [the specialized high schools] offer assist in developing the special talents of these exceptional students" (p. 91).

In Korea, specialized STEM schools were originally established by the effect of the High School Equalization Policy (HSEP) in 1980s. According to Choi and Hong[6], students were assigned to one high school in their school district by a lottery system. Random placement regardless of students' abilities implied that schools only focused on equalized setting for students in spite of individual students' various needs in their learning. This led to critics that many gifted students could not have education that could support their interests and needs. This resulted in the birth of Korea gifted education with the foundation of the first SHS, the Gyeonggi Science High School in 1983. SHSs had been the only type of specialized STEM schools before the appearance of the first Science Academy (SA), Korean Science Academy (KSA) founded in 2003 (read Choi & Hong[6] for more information).

1.2 *Secondary STEM Schools in Korea*

SHSs are public high schools for mathematically and scientifically gifted students that offer accelerated and enriched curricula for this special population compared to Korean national school curriculum. As they are public schools, SHSs have to follow the national curriculum as they are bounded by law. In practice, SHS students usually learn national curriculum in an accelerated manner and can learn advanced mathematics and science later years[6]. However, students and parents took advantage of accelerated education in SHSs as a college entrance exam preparation training to enter prestige universities, which was not the purpose of specialized STEM schools. And, such a practice resulted in the revision of Korea gifted education to maintain the purpose of STEM schools[16], which led to starting a new type of specialized schools, called SA.

SAs have more autonomy in designing their curriculum and select students regardless of school districts and ages of students[4,7,17]. Because of a success of the first SA - KSA for several years, it was decided to convert three more SHSs to SAs. To be specific, the Gyeonggi Science High School, the Seoul Science High School, and the Daegu Science High School have been approved to convert to SAs since 2008[a]. As of 2012, there are 23 specialized STEM schools including four SAs and 19 SHSs.

---

[a]Although the Seoul Science High School, the Gyeonggi Science High School, and the Daegu Science High School have converted from SHSs to SAs, the three schools have been using the original names including "Science High School (SHS)."

As the Korean government has great interests in and develop policies on various aspects of SHSs and SAs[20], there were a few reports on topics of mathematics or science teachers' perception about their curriculum, management, and early graduation in SHSs[17,22,27]. Oh's study[22] on the teachers' perception about curriculum and early graduation of SHSs emphasized a career intensive course implementation with a brief description of legal background of SAs and SHSs. Two others studies reported on specialized school teachers' perspective on curriculum and effects of early graduation of students[17,27]. However, there is no research investigating STEM or gifted school teachers, teacher training programs, teacher employment systems, or teacher qualification criteria for specialized STEM schools.

1.3 *Gifted Students and Their Teachers*

Feldhusen and Hansen[8] reviewed previous research on the characteristics of effective teachers for the gifted and talented students. Practitioners and university staff working for programs for the gifted generally indicated that the following are important; knowledge about nature and needs of gifted students, ability to develop creative problem solving skills, ability to develop methods and materials for use with gifted students, and skill in promoting higher cognitive thinking abilities and questioning techniques. Hansen and Feldhusen[14] compared teachers trained in gifted education to those without any training in gifted education to identify features of effective teachers for the gifted. They found that trained teachers made more positive class environments. Students learned from trained teachers inclined to higher-level thinking discussions while their counterparts paid more attention to lecture quality and grading systems.

When VanTassel-Baska and Johnsen[32] synthesized the previous studies on requirement for teaching the gifted, they stated what teachers of the gifted need to know; meta-cognitive strategies, higher level thinking strategies in content areas, and activities that address the gifted students' areas of interest. Chambelin and Chambelin[1] found that experienced teachers with gifted learners understood the needs to adapt instruction for this special group and highlighted the importance of student-centered instructions. While having broadened view about giftedness, these teachers attempted to align problem-solving tasks with students' interests and readiness.

Vialle and Tischler[33] compared students' perspectives on effective teachers for the gifted students across three countries - Australia, Austria, and the United States. Participated students showed fondness on teacher's personality traits over intellectual characteristics with a varied report on grade and gender preference. Vialle and Tischler's study described the similarities and differences in requirements for teachers of the gifted acorss the countries as well. For instance, in Russia, leading scienctists are hired as school teachers in two major types of specialized schools[11] while "gifted education in Europe is mainly education pursed in inclusive settings and is education signified by cultural variety"[23](p. 5). It will be informative and beneficial to learn about how specialized STEM schools in different countries recruit high-quality teachers for the gifted, as well as retain and improve the quality of teachers.

2   Data and Findings

2.1 *Data*

To collect information on teacher qualifications of secondary specialized STEM schools in Korea and what qualifications current STEM school teachers possess, we searched dissertations, academic journal articles published, secondary STEM schools' websites and reports published by the Korean government, public and private educational research institutions. We also tried to identify professional development programs that SAs and SHSs offer to retain and improve teachers' quality to teach gifted students.

Both types of specialized secondary STEM schools, SAs and SHSs, release annual reports, *school education plan,* to provide schools' plans for the next academic year as well as general information regarding curriculum, students, and teachers. For the year of 2012, three of four SAs and two of 19 SHSs posted their education plans on their website for public-access. In terms of professional development, the three SAs and two SHSs released information on their education plans including teacher professional development programs such as new teacher training, an open recruitment system, communication methods, and annual evaluation processes.

2.2 *Who Are Qualified to Teach at Specialized STEM Schools in Korea?*

Korea Foundation for the Advancement of Science and Creativity (KOFAC) is one of the educational research institutes and plays a significant role in supporting SAs and SHSs in developing and revising curriculum, improving admission process and teacher professional development programs. According to KOFAC[18], SAs and SHSs were different in their legal status (see the table II-1, KOFAC[18], p.37). SAs were established grounded on *Act on the Promotion of Specific Education for Brilliant Children (APSEBC)* supporting the gifted education only while SHSs are based on *Elementary and Secondary Education Act (ESEA)* which regulates public and private schools and teachers.

According to *ESEA,* all teachers in SHSs should be certified to teach and be transferred to another school every four or five years due to the standardization of schools. With a possibility of one-time extension of the employment contract, teachers in SHSs can teach in a SHS at most for ten years, and every teacher including the principal has to be a certified. On the other hand, SAs can hire teachers by "open recruitment system" that is not limited to certified teacher candidates. Particularly, those who have a doctoral degree without a teaching certification are encouraged to teach at SAs based on *APSEBC*. While teachers have limited years to teach in one SHS, SA teachers can extend their contracts with SAs as long as the teacher and the school decide to do so.

Seoul SA[28] (formally known as Seoul SHS) provided detailed information on desired qualifications of teachers candidates to teach the school's gifted students. Qualified teacher candidates for Seoul SA should have a record of successful teaching experiences in public schools and a recommendation letter from a public school principal. Non-certificated candidates, however, need a doctoral degree in mathematics or subfields of science to be considered to teach at the school. Also, teacher candidates in Seoul SA are expected to verify understanding and expertise in gifted education and ability to teach in English. However, teachers hired without teaching certificates are only considered as non-tenure track position. Considering that teachers without teaching certification are very likely to be advance degree holders, hiring them for temporary position causes difficulty to hire and retain teachers.

Candidates' qualifications are reviewed and screened through a multi-step process. The hiring process used by all SAs is open recruitment system[19] that begins with each SA's forming a personnel committee that conducts application document screening and interviews. Teacher candidates must demonstrate a teaching in front of the committee as a next step. As the final step, the principal of SA interviews the candidates to make a final decision.

2.3 *Current STEM School Teachers' Qualifications*

KOFAC[19] report provides details about current teachers in SAs. According to the report, 41.9% of teachers in KSA, the first SA, were certified to teach. On the other hand, 96.1%, 96.8% and 97.7% of teachers in Seoul SA, Gyeonggi SA (formally known as Gyonggi SHS), and Daegu SA (formally known as Daegu SHS) have teaching certification respectively. It may be due to the fact that the three SAs have been converted from SHS since 2008 without replacing existing teachers while KSA started as an SA, not an SHS. In other words, the majority of teachers in the three converted SAs had taught in previous SHS buildings where teachers were required to be certified to be hired. Table 6-1 shows numbers of teachers who have the teaching certification in each of four SAs. As seen in Table 6-2, over a half of KSA teachers possessed doctoral degrees while less than 20% of teachers the other SAs did. For the three SAs, more than half of teachers have a master's degree or higher.

Table 6-1. Numbers of Teachers with the Certification (KOFAC[19], p. 55)

	KSA	Seoul SA	Gyeonggi SA	Daegu SA
Principal	0/1	1/1	1/1	1/1
Vice principal	0/1	1/1	1/1	1/1
Teacher	26/62	74/77	60/62	42/43
Rate of teachers with the certification	41.9%	96.1%	96.8%	97.7%

Note: Numbers in the denominators indicate the total numbers of teachers in the category.

Table 6-2. SA Teachers' Terminal Degrees (KOFAC[19], p. 56)

	KSA	Seoul SA	Gyeonggi SA	Daegu SA
Doctoral degree	44	17	12	3
ABD	5	11	11	8
Master's degree	8	33	34	26
Master's completion	1	9	0	0
Bachelor's degree	6	9	7	8
Total	64	79	64	45

SAs ran teacher evaluation programs annually for purposes of providing useful information for teacher's development, sharing ideas and experience with peer teachers, and improving teachers' expertise to enhance quality of teachers and teaching[19]. Administrators, peer teachers, parents and students are engaged in the teacher evaluation process. They assess teachers' teaching and research activities and student guidance works in teacher evaluation[13,28]. The results of teacher evaluation are used to renew contracts and to develop professional development sessions for teachers (for details, see the Table II-50 KOFAC[19], p. 63).

SAs conduct various professional development activities to improve expertise. Generally, teachers in SAs can attend two types of teacher training programs; for research and for teaching methods. All SA teachers are encouraged to participate in research projects incorporating students who are interested in teachers' research topics. Professional development programs for teaching methods included learning pedagogical methods and developing instructional materials. Furthermore, teachers in SAs have regular meetings with each other and involved in a mentoring system to communicate and to share teaching and research ideas[19].

Among SAs, Seoul SA[28] provides the more detailed information about their professional development programs for teachers. They offer training programs for new teachers designed to help new teachers to understand and adapt to Seoul SA. Other programs include topics on learning innovative teaching methods for the gifted learners and developing teaching materials. SA's teacher mentor program intends to improve teachers' abilities through peer supervision and to promote discussion to share ideas. Both of collegial supervision and self-supervision help individual teachers contribute to reflecting and developing their own teaching methods. Seoul SA[28] emphasizes communication amongst teachers because it is believed as effective in further developing pedagogy and teacher evaluation methods. As teachers are encouraged to conduct research projects including their students in them, Seoul SA teachers offer research seminars as well. Other SAs encourage teachers to participate teachers-training program and research programs designed similar to those in Seoul SA.

Based on two SHSs which provided information about their teachers, they support teachers by similar ways to SAs such as teacher evaluation and self-supervision over their instruction. Gwangju SHS[12] and Sejong SHS[26] support teachers to conduct self-supervision over their instruction. Particularly, Gwangju SHS encourages teachers to open their classes to parents and peer teachers as a part of self-supervision while Sejong SHS recommends teachers to have a demo lesson. In Sejong SHS, teachers are encouraged to make learning groups among teachers to research on instructional innovative activities related to developing teaching materials and improving teaching ability.

3 Discussion

The importance and impact of teachers on students' learning and intellectual development, especially for the gifted learners, are widely recognized. To provide best possible education for the future STEM leaders of the country, specialized STEM schools are responsible to secure highly qualified and competent teachers. However, stakeholders do not have clear understanding of what expectations and qualifications are for teachers of such schools nor schools seem to have clear pictures of whom they want to hire for teachers of the gifted. As a starting point of this issue, we attempted to investigate what required qualifications for SA and SHS teachers are, how qualified current STEM school teachers are, and what opportunities current STEM school teachers have to further develop their qualifications. To learn about teacher qualifications (i.e., terminal degree, teaching experiences with general students as well as gifted students, etc.), we reviewed government documents, research reports, and school websites. This would be a good starting point to assess current situation of teacher quality of specialized STEM school and to prepare curriculum for future teachers of the gifted. This article focuses on teachers of specialized secondary STEM schools in Korea as Korean secondary school students' successful accomplishment in international comparison studies and International Mathematical and scientific Olympiads in recent years is of interests of many educators worldwide.

As two types of specialized secondary STEM schools in Korea — SAs and SHSs — are bounded by different legal regulations, expected qualifications for teacher candidates vary: SHS teachers should be certified to teach at secondary schools whereas SAs have more autonomy to hire teachers through that open recruitment system. New hires for SAs are very likely to possess advanced degrees including teachers without teaching certifications as well as of foreign nationalities. Learning from teachers with various backgrounds could prepare students to be global citizens as SAs want their students to become such. Teachers with advanced degrees in mathematics and science are often considered as good resources for students to explore advanced contents in STEM disciplines and to gain research experiences with the teachers. When there is a liberty to hire teachers with diverse backgrounds, in other words, when there are no fixed requirements for teacher candidates to have, the flexibility should be cautiously utilized in a way to meet effective teacher quality for the gifted students.

SAs and SHSs provide their teachers opportunities to participate in various professional development programs. To enhance previous teaching experiences and advanced degrees in their content disciplines, STEM schools offer professional development sessions, which could equip the teachers in a better position to teach gifted students in STEM. As the portion of teachers who are non-certified but having advanced degrees in the content disciplines increases in all SAs, it is predicted that teachers are not necessarily prepared to understand students and pedagogy. As it is noted that having advanced knowledge in the discipline is not a sufficient condition to be a good and effective teacher, teachers ought to be developed and focused to areas that teachers lack of specialties. Also, through various teacher evaluation systems, teachers receive supports and feedbacks from peer teachers, administration personnel, parents and students, which will help improve teaching future STEM leaders of the country. SAs' recruitment system, in-service teacher training programs, teacher mentor program, teacher evaluation, and research are all together to improve quality of STEM school teachers.

The fact that specialized STEM schools increase the numbers teachers with advance degrees to teach does not necessarily mean that mathematicians and scientists rave about working at SAs and SHSs. The employment policy of SAs for non-certified teachers seems somewhat odd that teachers who are not certified are only hired temporarily as non-certified teachers are likely to be advanced degree holders. With currently available information, it is not clear that what motivations these teachers work for unstable positions at STEM schools and how long they usually stay to teach at the schools. Certainly, there are rooms to improve to (1) find highly qualified teachers and (2) retain good teachers.

There are only limited expected or required teacher qualifications known. They are terminal degrees, teacher certification, and teaching experiences. Not all specialized STEM schools require their teacher candidates to have any background or knowledge to teach gifted students. As previous research asserted, teachers of the gifted are more effective and efficient when they are in rich knowledge in gifted education and trained to incorporate various teaching methods appropriate for the gifted learners. However, Korean STEM schools do not appear to require teacher candidates to have such qualification nor to further develop enrichment in the area. Rather, teachers are preferred to possess advanced degrees in STEM disciplines. As it is understood that Korean secondary schools are already successful in STEM subject achievements including gifted and general students, it could be a case that Korean students excel their performance and productivity in the future if teachers of specialized STEM schools are better prepared.

4 Conclusion

Main source of this investigative study is secondary data such as the government reports, published articles, and school websites. These restricted sources of data could only offer limit detailed information about teachers. Most SHSs did not open any information about teachers' activities while SAs make some details about teachers open to public. SHSs mainly provide sources regarding student admission process, their curriculum, and students' activities. It could be due to SAs' autonomy in hiring teachers while SHSs' hiring system is restricted by legal regulations. When a law prescribes what should be done in a certain way, there could not be much room to explain. This article may contribute to understand backgrounds and general ideas about who can be a teacher in SAs and SHSs and how they try to develop teaching abilities. However, further studies are needed in the areas of effects of teacher qualifications on gifted students' success, content of continuing professional development for STEM teachers for the gifted as well as general students, and teacher preparation programs for the gifted. This will influence on teacher preparation programs, especially, for the education of the gifted.

As examined schools assert, one purpose of specialized secondary STEM schools is to rear successful future leaders of the country in STEM fields. Students already have displayed their talents and interests in the STEM areas when they enter the schools, however, their further development are considerably influenced by their teachers. Advanced degrees in the discipline cannot be the only criterion for qualifications as previous research studies have widely disseminated that there is more than content knowledge of teachers that affect students learning and advancement. The schools need to enhance teacher quality in various aspects and curricula to provide best possible education for highly gifted and talented students in STEM fields.

References

1. Chambelin, M. T., & Chambelin, S. A. (2010). Enhancing preservie teacher development: Field experiences with gifted students. *Journal for the Education of the Gifted, 33*(3), 381–416.
2. Choi, K. (2009). *Characteristics of Korean International Mathematical Olympiad (IMO) winners' and various developmental influences.* (Doctoral disseration). Columbia University. ProQuest Dissertations & Theses database. (UMI No.3386133)
3. Choi, K. (2013). Influences of formal schooling on International Mathematical Olympiad winners from Korea. *Roeper Review, 35*(3), 187-196.
4. Choi, K. (2014). Opportunity to explore for gifted STEM students in Korea: From admissions criteria to curriculum. *Theory into Practice, 53*(1), 25–32.
5. Choi, K., Choi, T., & McAninch, M. (2012). A comparative investigation of the presence of psychological conditions in high achieving eighth graders from TIMSS 2007 mathematics. *ZDM: The International Journal on Mathematics Education, 44*, 189–199.
6. Choi, K., & Hong, D. (2009). Gifted education in Korea: Three Korean high schools for the mathematically gifted. *Gifted Child Today, 32*(2), 42–49.
7. Choi, K., & Jang, J. (2012). A recent history of Korean public institutions for the mathematically and scientifically gifted: From specialized science high schools to science academies. *Mathematical Creativity and Giftedness Newsletter, 3*, 15–18.
8. Feldhusen, J. F., & Hansen, J. B. (1988). Teachers of the gifted: preparation and supervision. *Gifted Education International, 5*, 84–89.
9. Fleischman, H. L., Hopstock, P. J., Pelczar, M. P., & Shelley, B. E. (2010). *Highlights from PISA 2009: Performance of U.S. 15-year-old students in reading, mathematics, and science literacy in an international context.* (NCES 2011-004). Washington, DC: The National Center for Eduational Statistics (NCES).
10. Flores, A. (2007). Examining disparities in mathematics education: Achievement gap or opportunity gap? *The High School Journal, 91*(1), 29-42.
11. Grigorenko, E. L., & Pamela, R. C. (1994). An inside view of gifted education in Russia. *Roeper Review, 16*(3), 167–171.
12. Gwangju Science High School. (2012). *2012 school education plan.* Gwangju, Korea: Author. [in Korean]

13. Gyeonggi Science High School. (2012). *2012 school education plan*. Gyeonggi, Korea: Author. [in Korean]
14. Hansen, J. B., & Feldhusen, J. F. (1994). Comparison of trained and untrained teachers of gifted students. *Gifted Child Quarterly, 38*(3), 115–121.
15. Hong, D. S., & Choi, K. (2014). A comparison of Korean and American secondary school textbooks: the case of quadratic equations. *Educational Studies in Mathematics*, 85, 241–263.
16. Kim, H. O. (1995, October). Are specialized high schools operated as its 'specialized purport'. *Joongdung Uri Gyoyook, 68*, 78–83. [in Korean]
17. Kim, Y. -S. (2012). *The Actual situation of science high school curriculum and of early graduation: In Dajeon-Chungnam area*. (Unpublished master's thesis). Korea National University of Education, Chung-Buk, Korea. [in Korean]
18. Korea Foundation for the Advancement of Science and Creativity [KOFAC]. (2009). *A study of educational development strategies to facilitate gifted education in science*. Seoul, Korea: Author. [in Korean]
19. KOFAC. (2012). *A study on the improvement of the gifted science high school*. Seoul, Korea: Author. [in Korean]
20. Ministry of Education, Science, and Technology. (2009). *Improvement plan for admission policy and system for High School advancement*. Seoul, Korea: Author. Retrieved from http://www.mest.go.kr/web/221/ko/board/download.do?boardSeqD36779 [in Korean]
21. Mullis, I. V. S., Martin, M. O., Foy, P., & Arora, A. (2012). *TIMSS 2011 international results in mathematics*. Chestnut Hill, MA: TIMSS & PIRLS International Study Center Lynch School of Education, Boston College.
22. Oh, S. E. (2010). *The research of science high school curriculum: The inquiry about possibility of career intensive course implementation by students' entering university*. (Unpublished master's thesis). Korea University, Seoul, Korea. [in Korean]
23. Rersson, R. (2009). Euroup, gifted education. In B. Kerr (Ed.), *Encyclopedia of giftedness, creativity, and talent* (Vol. 1). Thousand Oaks, CA: Sage.
24. Rogers, K. B. (2007). Lessons learned about educating the gifted and talented: A synthesis of the resaerch on educational practice. *Gifted Child Quarterly, 51*(4), 382–396.
25. Schack, G. D., & Starko, A. J. (1990). Identification of gifted students: An analysis of criteria preferred by preservice teachers, classroom teachers, and teachers of the gifted. *Journal for the Education of the Gifted, 13*(4), 346-363.
26. Sejong Science High School. (2012). *2012 school education plan*. Seoul, Korea: Author. [in Korean]
27. Seo, H. -A., Kwak, Y. S., Jung, H. –C., & Son, J. –W. (2007). Teachers' perceptions to management of science high schools. *Journal of the Society for the International Gifted in Science, 1*(2), 25–134.

28. Seoul Science High School. (2012). *2012 School education plan.* Seoul, Korea: Author. [in Korean]

29. Speer, N. M., & Wagner, J. F. (2009). Knowledge needed by a teacher to provide analytic scaffolding during undergraduate mathematics classroom discussion. *Journal for Research in Mathematics Education, 40*(5), 530–562.

30. Stephens, K. R. (1999). Residential math and science high schools: A closer look. *The Journal of Secondary Gifted Education, 10*(2), 85–92.

31. Thomas, J., & Williams, C. (2010). The histroy of specialized STEM schools and the formation and role of the NCSSSMST. *Roeper Review, 32*, 17–24.

32. VanTassel-Baska, J., & Johnsen, S. K. (2007). Teacher education standards for the field of gifted education: A vision of coherence for personnel preparation in the 21st centery. *Gifted Child Quarterly, 51*(2), 182–205.

33. Vialle, W., & Tischler, K. (2005). Teachers of the gifted: A comparison of students' perspectives in Australia, Austria, and the United States. *Gifted Education International, 19*, 173–181.

CHAPTER 7

MATHMATICS CAMP FOR MATHEMATICAL OLYMPIAD

Seunghun Yi

Sciences and Liberal Arts (Mathematics),
Youngdong University
Youngdong-eup, Youngdong-gun, Chungbuk 370-701, South Korea
E-mail: seunghun@yd.ac.kr

Among many other educational programs for the gifted students, the mathematics vacation camp program is considered as one of the best programs. The Korean Mathematical Olympiad (KMO) mathematics camp program plays a tremendous role in amplifying the mathematically gifted students' talents and it is one of the most important factors for South Korea to maintain its high rank in the IMO.

In this article I discussed two major topics: first, the structure of the mathematics camp of KMO and second, the emotional experience encountered by the participants of the mathematics camp.

KMO math camp is held twice a year. Summer class starts in August and winter class starts in January. Around 40 students are selected for each category of middle school and high school, and two classes are held separately for them. Classes are for 2 weeks and the students live together for the duration of the classes. The classes consist of lectures b professors, problem solving sessions by graduate students, and two practice tests.

During the math camp, the students socialized with one another, advanced academically through healthy competitions, taught one another through such interactions and received emotional support and consolation. They immersed themselves in mathematics, experiencing an overall increase in their mathematical skills. These experiences continue to affect them in a positive way even after the camp, leading them to be more confident and interested in mathematics so that they can ultimately major in mathematics in the future and choose to walk the path of a professional mathematician.

1   Introduction

In the 21st century knowledge-based society, countries all over the world have a great interest in gifted education and strive to identify the outstanding talents early in their development and train their abilities to secure high-quality human resources in order to strengthen the international competitiveness. South Korea initiated a full-fledged gifted education program in math and science with the establishment of the

Kyung-Ki Science High School in 1983 and has continued to expand the gifted education program.

Many countries have long been interested in Mathematics & Science Olympiads. It enabled them to discover and train many gifted individuals early in their development. Especially, Mathematical Olympiad is globally known as one of the most suitable ways to determine the mathematically gifted students and train their mathematical ability[2]. South Korea participated for the first time in the 29th International Mathematical Olympiad (IMO) held in Australia in 1988, and has continued to participate annually with excellent results, maintaining a high rank, even winning first place in 2012. With the support of the South Korean government, Korean Mathematical Society (KMS) is in charge of managing the Korean Mathematical Olympiad (KMO). Through KMO, KMS is discovering and training the gifted individuals early in their development, successfully cultivating a pool of highly educated individuals in science and engineering.

Among many other educational programs for the gifted students, especially the mathematics vacation camp programs are frequently being held in many countries throughout the world. South Korea focuses education for the gifted students by holding educational camp programs during the summer and winter vacations provided by Math and Science Olympiads and gifted development centers. It also utilize the educational camp programs when selecting new students for science high schools. Therefore camp programs are not only useful in educating gifted students but also in selecting qualified individuals for various schools. KMO mathematics camp program plays a tremendous role in increasing the interest in mathematics for the mathematically gifted students and amplifying their talents. Two major topics will be discussed in this paper. First, the structure of the mathematics camp, the core reason behind KMO producing excellent results in the International Mathematical Olympiad (IMO). Second, the emotional experience encountered by the participants of the mathematics camp.

2 South Korea's Participation of the IMO and the History of the KMO

South Korea participated in the 29th IMO in 1988 for the first time. In order to select Korean team that would participate in the IMO, KMS organized the Korean Mathematical Olympiad Committee (KMOC) in 1987 and conducted the 1st KMO. It selected 34 students who scored the highest and provided a 4-week long specialized education for the IMO during the 1st KMO Winter Camp in January 1988. Then, they conducted a Team Selection Test (TST) and chose 6 students to form Korean team. Korea participated in the 29th IMO in 1988 and took 22nd place (3 Bronze medals) out of 49 countries that participated. Since then, KMOC has continued to take charge and conduct KMO, and annually participate in IMO by selecting participants through KMO.

Korea has ranked in the top 20 in the initial stages, but took 4th place in the 41st IMO held in Korea in 2000. They maintained a high rank of 3rd or 4th place from then on and finally took 1st place overall in 2012 (see Table 7-1).

Table 7-1. Rank of Korean Team in the IMO

Year	Rank		Year	Rank		Year	Rank	
	Abs.	Rel.		Abs.	Rel.		Abs.	Rel.
1988	22	56.25%	1998	12	85.33%	2008	4	96.88%
1989	28	44.90%	1999	7	92.50%	2009	4	97.09%
1990	32	41.51%	2000	4	96.30%	2010	4	96.88%
1991	17	70.91%	2001	4	96.34%	2011	13	88.00%
1993	15	80.56%	2002	6	93.98%	2012	1	100.00%
1992	18	69.09%	2003	6	93.83%	2013	2	98.96%
1994	13	82.35%	2004	12	86.90%			
1995	7	91.67%	2005	5	95.56%			
1996	8	90.54%	2006	3	97.75%			
1997	11	87.65%	2007	3	97.83%			

* Abs.: Absolute rank, Rel.: Relative Rank

The initial purpose of the KMO was to select participants for the IMO, but as the Mathematical Olympiad continued to grow, it has expanded to discover the gifted students and train them early in their development. KMO is not just a simple competition for the IMO. It helps the elementary, middle school and high school students to amplify the mathematical interest, encourages them to study mathematics more diligently, and guides them to become an outstanding workforce further down the road in the field of natural sciences and engineering.

3   KMO's Annual Schedule

KMO annual program consists of two components: Test administration for purposes of selecting participants of IMO and education for the participants. The selection is done annually and is a 4-step process: KMO first round, second round, third round and final round. Each round grants the top scoring students to move on to the next round. The final round consists of 12 candidates, and top 6 students are selected following the final round after combining the scores for all 4 rounds.

Table 7-2. KMO's Annual Schedule

Date	Event
May – Jun	KMO 1st Round
Jul – Aug	Summer Camp
Sep – Oct	Online Train
Nov	KMO 2nd Round
Jan	Winter Camp
Feb – Mar	Online Train
Mar	KMO 3rd Round
Apr	Team Selection Test
	Korean Team

Education for the participants consists of two components: Math camp in summer and winter vacation and online training during regular semesters. Summer classes are in session during the months of July and August and Winter classes are in January, each session lasting about 2 weeks at a time. Online training targets graduates of summer and winter classes for 8 weeks. Class materials are mailed to the students weekly and students study independently at home. KMO's tentative annual schedule is outlined in Table 7-2.

4 Structure of the Math Camp

The Winter Camp in KMO has continued to be held since the 1st KMO Winter Camp in January of 1988. The distinguished mathematically gifted students that were recognized through the KMO first and second round have an opportunity to socialize and promote friendship, and challenge their mathematical abilities through healthy competitions. The increased mathematical skills and the ability to face challenges by participating in the KMO mathematical camp can motivate the participants to study harder, which can become one of the most important factors for South Korea to maintain its high rank in the IMO.

KMO math camp is held twice a year. Summer class starts in August and winter class starts in January. The participants for summer and winter classes are selected by their scores of the KMO first and second round. Around 40 students are selected for each category of Middle school and high school, and are held separately. Classes are for 2 weeks and the students live together for the duration of the classes.

Table 7-3. Whole Schedule of KMO Winter Camp

	Mon	Tue	Wen	Thu	Fri	Sat	Sun
Date			8	9	10	11	12
Event			Entrance	N	G	Exam	Exam
Date	13	14	15	16	17	18	19
Event	C	N	C	A	A	Exam	Exam
Date	20	21					
Event	G	Completion					

IMO covers Algebra (A), Combinatorics (C), Geometry (G), and Number Theory (N), and the students focus on solving mathematical problems in these four areas during the classes. The classes consist of lectures by professors who are experienced in IMO education, problem solving sessions by graduate students who have participated in the IMO or in KMO math camp, and two practice tests in the same format as the IMO are given on the weekend. The tentative schedule for the math camp students is listed in Table 7-3 and Table 7-4.

Table 7-4. Day Schedule of KMO Winter Camp

Time	Event
07:00 ~ 09:00	Breakfast
09:00 ~ 10:15	Problem Solving 1
10:30 ~ 11:45	Problem Solving 2
11:45 ~ 13:15	Lunch
13:15 ~ 14:30	Problem Solving 3
14:45 ~ 16:15	Lecture 1
16:30 ~ 18:00	Lecture 2
18:00 ~ 19:00	Dinner
19:00 ~	Free time

Generally, the top scoring students from first and second round are selected for the math camp. However, for the purpose of the selection of Korea Team for IMO, special consideration is given to girls, younger students, students from rural areas and students from regular high schools (non-science high schools). In other words, these students are given an opportunity to enter the math camp even if their KMO scores are lower.

Summer and winter classes are the same in that they both educate the mathematically gifted students, but winter classes are more geared towards selecting the participants for IMO and educating them, and summer classes are for discovering mathematically gifted students training them. Therefore, math camp students are selected according to the characteristics of each session with flexibility.

5   Students' Perception about the Math Camp

Let's examine the emotional changes of the mathematically gifted students who have been admitted to the math camp. The stages of the math camp can be divided into entrance, 1st practice examination, 2nd practice examination, and completion, based on the two practice examinations that are given on the weekends.

5.1 *Entrance*

The participants of the math camp were very proud to be chosen to participate in the camp and had great expectations filled with joy and excitement. Entering the math camp meant gaining an official approval that their mathematical skills are of the highest-caliber in South Korea. It also enabled them to meet other mathematically gifted students around the country and listen to the lectures of the best professors and the Graduate students.

5.2 *After the 1st Examination*

5.2.1 *Emotional Up and Down by Comparing with Others*

The 1st practice examination was given on the 4th and 5th day, the first weekend, of the math camp. The practice exam was 4 hours and 30 minutes long and included 8 descriptive problems that were of IMO level of difficulties.

The 270-minute-long practice exam with 8 problems of IMO level difficulties is a very challenging exam even for the mathematically gifted students. They concentrate for hours and make every effort to solve the challenging problems. They are under a lot of stress from concentrating for a long time, but they encouragingly accept the challenging process and demonstrate strong commitment for the task.

The result of the 1st practice examination influences the students a lot. The students can become very disappointed if the results are low, but good results can give them a confidence boost. They were more strongly influenced by the relative scores from comparing with others than the raw score itself.

5.2.2 *Trying to Exert*

The students can be disappointed if the exam results were relatively lower than the others, and may even feel that they lack the mathematical ability. But instead of being discouraged, they tried to exert themselves to overcome their setback and frustration.

One can assume that the students' failure tolerance is high through their pushing themselves to do better after experiencing a setback. Failure tolerance is the tendency to react in a relatively constructive way about a failure[3]. Those with high failure tolerance react in a more positive and constructive way than those with low failure tolerance[1,3]. They have a high tendency to demonstrate future-oriented aspect of behavior even in unfavorable situations such as failure without dwelling in the counteractive state of mind, become more interested in what future has in store for them and take effective measures to academic success.

There was no difference in the results of the students' self assessment of their mathematical skills and ability between before their participation of the math camp and after the 1st practice examination. The fact that the results of their self assessment of mathematical skills and ability did not change after being disappointed by the relatively lower practice examination scores meant that their self assessment of their skills and ability was stable. One can assume that such stability was closely linked to failure tolerance as well as a sense of self-efficacy.

5.3 *Between 1st Exam and 2nd Exam*

5.3.1 *Improvement upon Competition*

The mathematically gifted students positively rated studying in a competitive environment by ability grouping. They demonstrated the tendency to strive to improve themselves as they study with other mathematically gifted colleagues.

The students were highly motivated to study harder as they interacted with other brilliant colleagues. They demonstrated the tendency to make a resolution to try harder in order to further increase their ability and implemented it, rather than being intimidated by other brilliant colleagues. One can assume that the reason behind the increased motivation and effort in a competitive environment was positively influenced by the students' firm and stable belief of their mathematical talent. Also, the fact that it was more effective for them to study in a competitive environment with other talented students supports the research that the ability grouping plays a positive role in the mathematically gifted students' scholastic achievement[4,5,6,7,8].

The students helped one another while studying together in the math camp. The person that was the most helpful while they were studying for the Mathematical Olympiads was a "friend". For them, a friend is a competitor as well as an ally to teach one another.

Through studying together with the peers who are similar to their scholastic level, they learn what they could not learn when studying by themselves. They were able to further increase their ability by being motivated to try harder with a strong sense of rivalry triggered by studying with others and by learning their expertise. These advantages might be the reasons why the mathematically gifted students favor the ability grouping.

5.3.2 *Determination to Make Up for the Failure*

The 2nd practice examination was given one week after the 1st practice examination. The students who were disappointed by the relatively lower scores of the 1st practice exam but made a resolution to try harder in order to make up for the setback have tried their best to study during the lectures, problem solving sessions and self-study sessions.

They were determined to obtain successful results in the 2nd practice examination in order to make up for the failure in the previous exam and studied hard during the classes. But there were various emotional responses about the 2nd practice exam. Those who prepared well during the week and were anticipating good results in the 2nd practice exam showed a combination of fear and excitement. Those who prepared well but did not get satisfying results in quizzes during classes and were not expecting as good results as they hoped were dreading the 2nd practice exam, and those who received somewhat satisfying results in the 1st practice exam and have been preparing even more for the 2nd practice exam showed more excitement than fear. As discussed above, the emotional aspects of the 2nd practice examination varied between students depending on their results of the 1st practice examination and the results of their performance during the preparation process thereafter.

5.4 *2nd Examination*

As with the 1st practice examination, their emotions were negatively influenced by the relatively low scores of the 2nd practice exam and positively influenced by the relatively high scores. But there was no difference in the results of the students' self assessment of their mathematical talents and ability between before and after the 2nd practice exam.

5.5 *Closing*

The students felt both pride and sadness to leave as the mathematical camp of 15 nights and 16 days came to an end. They were thrilled to have spent time with the other mathematically talented students, focused on studying mathematics and successfully completing the program. They reflected on the fact that they could have tried harder during the course of the camp, but also satisfied to see that there was an improvement on their mathematical abilities. Some reported an improvement not only on the academic aspects of themselves but also on the affective aspects such as self-control and sense of independence during the mathematical camp.

6   Conclusion

Two major topics were explored in detail in this paper. The first one was the structure of the Math Camp, the core reason behind KMO producing excellent results in the IMO. The second one was the emotional experience encountered by the participants of the mathematics camp.

During the Math Camp, the students socialized with one another, advanced academically through healthy competitions, taught one another through such interactions and received emotional support and consolation. They immersed themselves in the world of mathematics during the two-week camp period, experiencing an overall increase in their mathematical skills such as an increase in problem solving abilities and developing a mathematical insight. These experiences continue to affect them in a positive way even after the camp, leading them to be more confident and interested in mathematics so that they can ultimately major in mathematics in the future and choose to walk the path of a professional mathematician.

The following lists are some of the suggestions to improve the Math Camp. The mathematically gifted students rated studying in a competitive environment in their favor. But if we take into account that the students are emotionally sensitive teenagers and that some research suggests that competition may negatively affect the gifted students, careful consideration is deemed necessary regarding their emotional changes. Due to the fact that the students' emotional changes were influenced greatly by the relative comparison with other students, it is necessary to formulate a plan to lead them to focus on developing their ability through mastery.

Also, there is a need to expand their bond and friendship formed during the mathematics camp. The students reported that the most helpful person while they were studying for the Mathematical Olympiad was a "friend". They are fully aware of the advantage of studying as a group that increases their abilities as they learn the strengths from each other. Therefore it is necessary to provide as many opportunities as possible to study together and bond with one another during the camp. It would also be beneficial to help them extend the friendship even after the camp is over. A plan is necessary to help them interact with one another after the camp either on-line or off-line.

References

1. Clifford, M. M., Kim, A., & MacDonald, B. A. (1998). Responds to failure as influenced by task attribution, outcome attribution, and failure tolerance. *Journal of Experimental Education*, 57, 9-37.
2. Karp, A. (2003). Thirty years after: The lives of former winners of Mathematical Olympiads. *Roeper Review, 25*(2), 83-87.
3. Kim, A., & Clifford, M. M. (1988). Goal Source, goal difficulty, and individual difference variables as predictors of response to failure. *British Journal of Educational Psychology*, 58, 28-43.

4. Kulik, J. A. (1991). *Ability grouping and gifted students.* Invited address presented at the National Research Symposium on Talented Development, University of Iowa, Iowa City, IA.

5. Kulik, J. A. (1992). *An analysis of the research on ability grouping: Historical and contemporary perspectives. Storrs*, CT: National Research Center on the Gifted and Talented, University of Connecticut.

6. Kulik, J. A., & Kulik, C.-L. (1991). Ability grouping and gifted students. In N. Colangello and G. A. Davis (Eds.), *Handbook of gifted education* (pp.178-196). Needham Heights, MA; Allyn and Bacon.

7. Rogers, K. B. (1991). *The relationship of grouping practices to the education of gifted and talented learners.* CT: National Research Center on the Gifted and Talented, University of Connecticut.

8. Stanley, J. C. (1991). An academic model for educating the mathematically talented. *Gifted Child Quarterly*, 35, 36-42.

CHAPTER 8

USE OF TECHNOLOGY IN SECONDARY MATHEMATICS EDUCATION IN KOREA

Son, Hong Chan

Department of Mathematics Education, Chonbuk National University
#567 Baekje-daero, Deokjin-gu, Jeonju-si, Jeollabuk-do, 561-756,
Republic of Korea
E-mail:hcson@jbnu.ac.kr

This chapter is devoted to a survey of technology use in secondary mathematics education in Korea. The transition of technology use in secondary mathematics curricula and state of technology use in textbooks were observed, in addition to characteristics of in-service teacher training in relation to technology use.

1 Transition of Technology Use in Secondary Mathematics Curricula of Korea

The Korean government establishes and decides the national curriculum for the national education system. Primary and secondary school curriculum is organized and applied according to 'general curriculum' announced in a document from the government and has broad impact, not only on the writing of textbooks but also on teaching methods and school evaluations. In the case of mathematics, subjects are selected and then content, scope and units of completion are decided. Korean mathematics curriculum has been revised numerous times since 1946, with one revised in 2009[a].

Curriculum has been repeatedly revised to reflect global influence and trends. Emphasis on rigorousness of mathematics was implemented in the third mathematics curriculum released in 1973, due to NEW MATH

---

[a] The general curriculum of 2009 revision was released in 2009, and mathematics curriculum was released in 2011.

Movement. From the fourth curriculum, the focus has been on enhancement of mathematical problem solving ability. The seventh curriculum and 2007 revision give attention to the development of mathematical power and advanced thought, while the 2009 revision is focused on creativity.

The advancement of technology has significant influence on mathematics curriculum. In 1989, the National Council of Teachers of Mathematics (NCTM) in the United States introduced a new orientation to the computers and calculators to math curriculum and evaluation, stating that their potential to influence math education and encouraging widespread use in teaching and learning[10]. In 2000, the NCTM said: "Technology is essential in teaching and learning mathematics; it influences the mathematics that is taught and enhances students' learning" in the technology principle, which is one of six principles of school mathematics[11]. Many countries worldwide recognize the importance of technology and computers in the teaching and learning of mathematics, as they encourage active use of technology in education.

Korea has also advocated the use of technology since 1992, in response to global trends and modern technological development[6]. The statement on technology use in the sixth mathematics curriculum in describing the 'method of teaching and learning mathematics' is as following:

> "In order to perform complicated calculations and to improve problem solving ability, a calculator or a computer may be used in teaching and learning mathematics." (p. 42)

By the time when the sixth mathematics curriculum was established, 16-bit personal computers were getting into wide uses and software for mathematical education was not as broad or diverse as today's options. As a result, calculators and computers are indicated as tools for complicated calculation or improvement of problem solving ability, focusing on the functional aspect as auxiliary means.

In the seventh mathematics curriculum released in 1997, the use of technology like calculators and computers is mentioned:

> What follows should be considered in instruction of mathematics for the national common basic curriculum.

(1) Various forms of educational technology should be actively utilized.
(2) Except for those contents in which the enhancement of computational skill is primary, calculators and computers should be actively utilized to improve the understanding of concepts, principles, and rules, and to enhance problem solving abilities. (p.86)

It indicates that in the seventh curriculum, not only the problem solving, but also the aspects of understanding mathematical concepts, principles and laws are suggested for active use[7].

In the mathematics curriculum revised in 2007, the use of educational material in 'method of teaching and learning mathematics' and 'evaluation' is stated respectively as following:

In the use of educational materials in the teaching and learning of mathematics, the following must be kept in mind.
(1) Improve the effect of mathematical education by utilizing proper and various educational materials throughout the whole educational process of teaching and learning.
(2) In units that are not aiming for improvement of calculation abilities, technology like calculator, computer or educational software may be used to perform complicated calculations and understand mathematical concepts, principles and laws and enhance problem solving abilities. (p. 36)

In the evaluation of mathematical learning, students may be provided with the opportunity to use technology such as the calculator and computer, depending on the studied units. (p. 37)

What is stated in the 2007 revision of the curriculum is very similar to the relevant content in the seventh curriculum, but the 2007 revision also mentions the use of educational software and utilization of technology in evaluation, not only in teaching and learning methods[8]. Current math curriculum, the 2009 revision, has a similar stance to 2007, mentioning the use of technology in 'teaching and learning' and in 'evaluation', as well as the possible use of educational software for 'evaluation'.

Korean mathematics curriculum has consistently emphasized and broadened the effective use of technology in the sixth, seventh, 2007 and 2009 revisions.

2 Transition to Use of Technology in Secondary Mathematics Textbooks

All textbooks in Korea are influenced by educational curriculum, which is applied at the national level. As explained previously, Korean mathematics curriculum has emphasized the usefulness of technology in the sixth, the seventh, 2007 and 2009 revisions, and expanded the range of use. However, actual use of technology in secondary mathematics textbooks is comparatively rare and less present[1,3]. Lee argued that textbooks from 13 different publishers for 5 high school mathematical subjects do not sufficiently emphasize the use of technology and added that it is difficult to actually complete exercises with tools due to limited availability of technology[3]. In contrast, Kim and Son[2] show how technology has been applied and changed in textbooks of secondary mathematics curriculum in Korea. Kim and Son investigated 180 textbooks used in the sixth, seventh and 2007 revisions, then analyzed content areas technology used, types of technology and methods of using technology in teaching and learning mathematics.

First, Table 8-1 summarizes results from analysis of 180 textbooks in four content areas: algebra, analysis, geometry, and probability and statistics.

Table 8-1. Percentage of Technology Use by Content Areas. Frequency (percentage)

curriculum / content area	6th	7th	2007 revision	total
algebra	12(48%)	68(27%)	48(14%)	128(20%)
analysis	9(36%)	69(28%)	167(47%)	245(40%)
geometry	1(4%)	70(28%)	86(24%)	157(25%)
Probability and statistics	3(12%)	41(17%)	54(15%)	98(15%)
total	25(100%)	248(100%)	355(100%)	628(100%)

Except for textbooks of the sixth curriculum, in which the algebra and analysis took the largest proportions, the rest has comparatively uniform distribution of the use of technological tool and programs. Overall, analysis content logged the highest proportion at 40% and probability and statistics the smallest at 15%.

Looking at technology and program use in each revision of curriculum, frequency has increased from 25 to 248 and 355 in the sixth, seventh and 2007 revisions, respectively. An increase in frequency was significant between the sixth and seventh curriculum.

The limited use of technological tools in the sixth curriculum may be explained by the time period, as it was 1992. The lack of technological development and the fact that it was the first curriculum to mention technology use could have also been factors.

A diagram shows a distribution comparison of technology use between curriculums (Figure 8-1).

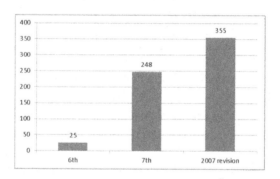

Figure 8-1. Use of Technology in the Sixth, Seventh and 2007 Revision.

In the sixth curriculum, the use of technology was more prominent in algebra (48%) and analysis (36%). In contrast, the seventh had an even distribution of use in each content area, which may be attributed to consistent insertion of technological tool exercises at the end of each chapter or topic. In the 2007 revision, analysis (47%) accounted for the largest proportion. Other content areas encouraged technology use, but a variety of graphing software made it easier to present in analysis content.

Second, the type of technology and programs were classified as: personal computer (PC), calculator and graphical calculator (GC), symbol manipulation and graphing software (SG), dynamic geometry (DG), spreadsheet (SS), program language (PL), internet (I) and others (E) (Table 8-2).

Table 8-2. Distribution of Technology Type Used in Curricular Frequency (percentage)

technology \ Curriculum	6th	7th	2007 Revision	total
PC	1(4%)	3(1%)	8(2%)	12(2%)
GC	15(60%)	57(23%)	43(12%)	115(18%)
SG	1(4%)	44(18%)	139(39%)	184(29%)
DG	0(0%)	46(19%)	66(19%)	112(18%)
SS	0(0%)	27(11%)	47(13%)	74(12%)
PL	8(32%)	2(1%)	4(1%)	14(2%)
I	0(0%)	68(27%)	13(4%)	81(13%)
E	0(0%)	1(0%)	35(10%)	36(6%)
total	25(100%)	248(100%)	355(100%)	628(100%)

Calculators and graphical calculators and programing languages are the main technology used in the sixth curriculum textbooks. The graphical calculator was the most popular tool, suggesting that the sixth curriculum was the first to contain the use of calculators and computers in 1992.

However, by the seventh curriculum, a more diverse use of technology and programs was introduced. Calculators and graphical calculators were introduced in algebra, graphical calculators and graphing software (SG) such as GrafEq in analysis, dynamic geometry software (GSP) in geometry, and spreadsheets in probability and statistics. Of note, the Internet was implemented evenly in all content areas, as a result of widespread use.

Similar to the seventh curriculum, the 2007 revision used varied types of technology. Graphing software (SG) in analysis was found to be the type of technology used most often. Attempts were made to use a diversity of tools in educational materials, namely GC in algebra, SG in analysis, DG in geometry and SS in probabilities and statistics. However, internet use significantly declined, and few new programs were introduced.

Last, the methods of using technology in mathematics textbooks were observed. Kim and Son[2] classified methods as: suggesting information on technology, explaining functions of technology, using technology as auxiliary equipment, using technology in guided investigation.

According to Kim and Son, the method was slanted toward explaining functions of technology and using technology as auxiliary materials in the sixth curriculum. The proportion of suggesting information on technology, explaining functions of technology and using technology as auxiliary equipment significantly increased in the seventh revision. Suggesting information on technology showed a sharp drop in the 2007 version. Using technology in guided investigation began to increase by the seventh curriculum and significantly increased in the 2007 revision, indicating that students were encouraged to understand and investigate mathematical properties and laws with the help of technology.

Analysis shows that the use of technology and mathematical software became more frequent each time the curriculum was revised, expanding use in all areas. Further, the method of using technology has been diversified, especially towards the method in which students can explore and learn mathematical properties with the help of technology. Secondary mathematics textbooks for the 2009 curriculum, currently applied in year 1 and 2 in middle school and year 1 in high school, are expected follow the same trend, as the revision continues to emphasize the importance of technological use.

3   Use of Technology in Secondary Mathematics Teacher Training

In the meaningful and effective use of technology in teaching and learning of mathematics, the knowledge and attitude of teachers on technology use are important.

It has been a long time since technology use in mathematics education was first introduced in 1992, and now more than 89.7% of all colleges of education in Korea offer lectures on the technology use in mathematics education[12]. With the emphasis on importance of technology use in Korean mathematics education, one may assume that the study of technology use among pre-service teachers has also increased.

However, analysis on the actual state of in-service teacher knowledge and attitude on technology use revealed a lack of education on technology use among pre-service teachers and suggested that experienced teachers were less proactive in utilizing technology than less experienced teachers.

Lee and Kim[4] attempted to understand the real state of mathematics teacher knowledge and attitudes toward technology use by observing 131 secondary mathematics teachers with credits from in-service training courses for 1st grade mathematics teacher in education offices of 16 cities and provinces in Korea.

According to Lee and Kim, use of technology was taught in most of in-service training courses of 16 education offices. However, it appeared that only 5.6% of teachers attended training use technology actively in teaching and learning mathematics, and 21.1% rarely utilize it and 30.6% use it in special cases. This suggests that most teachers are not using technology in their lessons.

Further, of those asked whether training time on technology was adequate, 27.8% answered positively and 69.4% answered negatively. Results indicate that the majority only have superficial knowledge on the basics of using technology. More than 90% of in-service teachers studied 1-3 technological tools in training, with the largest proportion 76.9% learning GSP, 70.4% Excel and 39.8% Cabri. Poly, Grafeq, Fathom and Winplot were classified as 'other' since none reached 4% individually. In summary, GSP, Excel and Cabri are the most studied tools in in-service teacher training courses in Korea.

A secondary mathematics teacher in Korea is typically entitled to attend in-service teacher training course for 1st grade mathematics teacher after 3 years. However, results above suggest that less experienced teachers rarely use technology in their lessons and that technology training is insufficient. It can be inferred that the study of tools lacks diversity, as few are being taught. Further, these results also highlight a shortfall of technology use at the college of education.

Experienced teachers were more likely to attend the duty training course rather than the qualification training course for 1st grade teachers, which is only done once. However, experienced teachers had a tendency to be less proactive in using technology than less experienced teachers because duty training courses did not sufficiently cover the use of technology. Therefore, the overall use of technology in mathematics lessons in Korea is said to be inactive.

4   Closing Remarks

Technology can help students learn mathematics by illustrating and complementing abstract and formal mathematics with accurate calculations and visual aids. It assists students with problem solving, reasoning and communicating, which allow development of mathematical thought, a critical objective of mathematical education. Further, the use of technology enables project or research study, which can significantly aid students in forming cooperation and acquiring knowledge[5].

Korean secondary mathematics curriculum has been actively encouraging technology in teaching and learning mathematics since the sixth curriculum, with widespread agreement on its benefits. It has also been shown that the use of technology is more actively applied in modern textbooks compared to previous editions. However, a survey of teachers attending in-service training course suggests that the use of technology in actual teaching and learning mathematics in school is not active.

In addition to the degree of technology used in mathematics curriculum and textbooks, and the knowledge and attitudes of teachers on technology use, the state of physical and social environments is also an important factor in using technology in teaching and learning mathematics.

Park et al.[12] reported that Korean students appeared able to complete routine tasks with computers, but confidence is lacking with difficult tasks. Further, Korean high school students face extreme pressure of university entrance examinations, and it is difficult to incorporate technology in high school lessons in which students are expected to solve lots of problems to score high. And pre-service teachers are unlikely to invest time in studying the use of technology since technology use is not a subject covered by the highly competitive teacher certification examination. Limitations are unlikely to be resolved in near future in Korea.

However, mathematics curriculum and textbooks aim for active use of technology in the classroom, and research on the use of technology is spreading. Teacher-centred workshops relevant to technology have become popular. Thus, the use of technology in Korean secondary mathematics education is expected to show a gradual increase in active participation and implementation.

References

1. Kim, M. (2005). *The analysis of the characteristics of Korean mathematics textbooks and "Contemporary Mathematics in Context" in USA in the viewpoint of Technology.* Unpublished Master's Thesis at Korea National University of Education. [in Korean]
2. Kim, M. & Son, H. (2013). The analysis on utilization trend of the technology in secondary mathematics textbooks based on the 6th, 7th, and 2007 revised curriculum in Korea. *Journal of Korean Society Educational Studies in Mathematics,* 15(4), 975-994. [in Korean]
3. Lee, J. (2010). A study on the use of technology in teaching-learning school Mathematics. *Communications of Mathematical Education,* 24(1), 29-48. [in Korean]
4. Lee, J., & Kim, Y. (2013). An analysis of the secondary school mathematics teachers' in-service training relating to the technology. *KNUE Journal of Mathematics Education,* 5(2), 95-108. [in Korean]
5. McGhee, R., & Kozma, R. (2003). New teacher and student roles in the technology-supported classroom. *Annual Meeting of the American Educational Research Association,* Seattle, WA.
6. Mistry of Education (1992). *The 6th middle school curriculum of the republic of Korea,* Seoul, Korea: Author. [in Korean]
7. Ministry of Education (1997). *The 7th mathematics curriculum.* Seoul, Korea: Author. [in Korean]
8. Ministry of Education and Human Resources Development (2007). *Mathematics curriculum revised in 2007.* Seoul, Korea: Author. [in Korean]
9. Ministry of Education, Science and Technology (2009). *Mathematics curriculum revised in 2009.* Seoul, Korea: Author. [in Korean]
10. National Council of Teachers of Mathematics (1989). *Curriculum and evaluation standards for school mathematics.* Reston, VA: NCTM Press.
11. National Council of Teachers of Mathematics (2000). *Principles and standards for school mathematics.* Reston, VA: NCTM Press.
12. Park, K., Jung, Y., Kim, W., Kim, D., Choi, S. & Choi, J. (2010). *A research on the developmental plan for mathematics education in elementary and secondary school.*

Seoul: Ministry of Education, Science and Technology & Korea Foundation for the Sciences and Creativity. [in Korean]

CHAPTER 9

DEVELOPMENT OF MATHEMATICS EDUCATION IN KOREA: THE ROLE OF THE KOREAN SOCIETY OF MATHEMATICAL EDUCATION

Inki Han

Department of Mathematics Education, Gyeongsang National University
501 Jinjudaero, Jinju City, Gyeongsangnam-Do, 660-701, Korea
E-mail: inkiski@gnu.ac.kr

Dohyoung Ryang

Department of Mathematics & Statistics
The University of North Caroline at Greensboro
#110, 116 Petty Building, 317 College Ave., Greensboro, NC 27412, USA
E-mail: d_ryang@uncg.edu

Jinho Kim

Department of Mathematics Education, Daegu National University of Education
219 Jungang-daero, Namgu, Daegu City, 705-715, South Korea
E-mail: jk478kim@dnue.ac.kr

The role of the Korean Society of Mathematical Education has been central in the development of mathematics education in Korea for the past half century. The present article reviews how and what the society has worked for the development of mathematics education of Korea. In summary, the society has contributed to collecting and analyzing information on the trends of international mathematics education as well as evaluating and suggesting reforms of mathematics curriculum of Korea. Also, the society publishes journals and books, and hosts national and international conferences. The present article is a reflective and evaluative review, in a historical view, through which we can see a future way of mathematics education of Korea.

1 Introduction

An excellent scholar greatly affects the development of his or her area of study. However, the actual progress in an academic area is accomplished by way of many other scholars in a scientific community continuously sharing and developing new ideas and creative approaches in the area. In other words, a science has been developed by their accumulating objective knowledge. Indeed, an individual scholar may first present his or her ideas to their peers. Then, other scholars cross-check and evaluate the scholar's subjective ideas to reconstruct these ideas into objective knowledge which is more acceptable to others. In general, a scholar presents his or her own ideas at an academic conference or in a journal, or personally communicates with other scholars. As observed, mathematics has been developed in this way for many centuries.

Teaching and learning mathematics does have an equally long history as mathematics itself does. In Korea, by an ancient record, mathematics was taught in *Gookhak* which was a kind of higher education institute founded in 682 during the Silla Dynasty[9]. Then, professors in *Gookhak* might think about how to teach mathematics effectively. This practical way of mathematics education still continues today. On the other hand, mathematics education as a science does not have a long history. A formal and theoretical approach to mathematics education has developed in modern times. Many scholars' integrated efforts through academic organizations, societies, and journals have actually contributed to the development of mathematics education. As results of their efforts, now mathematics education has become independent from psychology or education only recently[8]. In Korea, since the late 1960s, mathematics education community including mathematics education researchers, societies, and academic journals dealing with theoretical approach to mathematics education has been formed as supported by Korea Research Foundation and Korea Foundation for the Advancement of Science and Creativity.

What is mathematics education as a science? This question is not easy to answer. From when exactly was mathematics education created as a science in Korea? This is also difficult to answer. One key to answering these questions is to examine the activities of societies relating to mathematics education. These societies are playing critical roles, sometimes cooperating with each other and other times working apart, in developing the science of mathematics education. This work considers the Korean Society of Mathematical Education, the first academic society to study mathematics education in Korea, and recognizes its role in developing the mathematics education of Korea by analyzing articles published in the journals of the society.

2   Korean Society of Mathematical Education

2.1 *Establishing the Korean Society of Mathematical Education*

There are several academic societies in Korea in regard to research on mathematics education such as the Korean Society of Mathematical Education (KSME), Korea Society of Educational Studies in Mathematics (KSESM), the Korean School Mathematics Society (KSMS), and Korea Society of Elementary Mathematics Education (KSEME). The KSME was established in 1962, the KSESM in 1991, KSSM in 1998, and KSEME in 1996.

Let us see historical backdrop in which the KSME was established. Korea was independent from Japanese imperialism in 1945, August 15. Schools started at October in 1945, but many difficulties rose up because of no curriculum and textbooks. Choe[1] stated that "Japan was defeated in the World War II and the country was freed from Japanese imperialism, and then schools opened the door. Thus, we ought to do mathematics education by ourselves. However, due to the lack of both teachers and preparedness of textbooks, the education system fell into confusion again. Most schools still used Japanese textbooks written before the Liberation" (p. 22). In other words, right after the Liberation, the country did not have the human and material resources required, thus had no choice but to use textbooks and other materials used from before the Liberation in the teaching of mathematics.

Since then, in 1946, the Department of Education of the U.S. Army Military Government in Korea (USAMGIK) published the mathematics syllabus, and according to the syllabus, mathematics textbooks were developed and distributed. However, Choe[1] pointed out that "the syllabus looks different from that of Japan. But, the content is not much different from the one used before the Liberation (1942–1945), and the textbooks also follow the Japanese style in content and in writing format" (p. 22). To make matters worse, the Korean War broke out in 1950 and so revising the syllabus must be delayed. In conclusion, Korea did not have a choice but to follow Japan's mathematics education system even after the Liberation up until 1955 when the first national mathematics curriculum (MC-1) was published after the war.

The first national mathematics curriculum (MC-1) published in 1955 had been characterized by the real-life unit learning. Park[15] stated that "In June of the year 1950, the department of Culture and Education organized a council for establishing the syllabus and was about to review the syllabus leading up to the Korean War ... Since the government moved to Busan in the middle of the war, many educators residing in Busan participated in the council. They were then centered in the new curriculum movement with reference to Japanese literature, which was the real-life unit learning" (pp. 56-57). The MC-1 had been prepared for its development during the Korean War.

It is not expected that the MC-1, developed in the vortex of the war, had collected various opinions from mathematics educators and researchers as well as reflect students' developmental characteristics. In those days, Dewey's pragmatism had been introduced via Japan to Korean educators and had a big effect on MC-1. Also, "a U. S. educational mission came to the Country and guided the real-life unit learning" (Park[15], p. 60). In conclusion, the MC-1 is not only developed from necessity such as social demands or research results, but it was also developed by following the trends in mathematics education of foreign countries. Lee[11] pointed out that: "the mathematics textbooks developed on the basis of the MC-1 period had characterization putting on the real-life unit learning but its content did not actually follow the guide well, since mathematics textbook writers did not fully understand the real-life unit learning and an educational reform in the U. S. had already begun to move toward the discipline-centered curriculum," (Lee[11], pp. 43-44).

Many issues had occurred while implementing the MC-1. The Department of Culture and Education[3] evaluated the MC-1 by stating that "We, after the Liberation, established an education curriculum by ourselves for the first time. However, we did not possess the ability to implement the curriculum even with the collected reference materials. We never had experimental and survey data from a school, could not set goals in each subject taught in elementary, middle, and high school, and in fact, the MC-1 is, straightly speaking, written by guessing and imitating" (p. 2). Therefore, since 1958, the Department of Culture and Education had collected basic data for revising the MC-1 such as reports on implementing the MC-1, curriculums of other countries, curriculum principles, and UNESCO materials. To reflect the MC-1 and to prepare the second mathematics curriculum, the government and mathematics educators collectively would analyze the collected data, find the factors to mathematics education, and find a right way for advancing mathematics education. After all, it would be heartily required for them to collect and analyze systematic information to international trends of mathematics education at that time, seek and find a right development orientation to mathematics education, and conduct systematic researches on a variety of variables of mathematics education. Such demands had stimulated leading mathematics educators and researchers to make a group to represent the academic world of mathematics education in Korea so they can respond to such demands more efficiently in an official manner. This effort leaded to establish the KSME in 1962. Park[16], who took the lead in establishing the KSME, discussed:

> In the days of confusion, after the Liberation, the teaching syllabus which is the foundation for school mathematics was announced in 1946, and after that, the MC-1 was announced in 1955 after the Korean War, which imitated the curriculum of the neighbor country with no investigation or research. However, the imitation was blind not based on their own full researches and reviews. Just in time, the New Math movement rose up in the U. S. which influenced Japanese education in the name of Mathematics Education Modernization. However, not knowing the international trends in mathematics education, the government had prepared to revise the MC-1 for the second mathematics curriculum in entering 1960s. In this circumstance, mathematics educators, who agreed to share information on international mathematics education and encouraged conducting research studies on mathematics education, came together and established the KSME (p. 1).

In conclusion, the KSME had been established with the goals and roles of (a) collecting, analyzing, and disseminating information on international trends of mathematics education, (b) encouraging research studies on mathematics education, (c) meeting the demands of the times in mathematics education for Koreans (ourselves), and (d) improving mathematics classrooms.

2.2 Development of the KSME

The KSME had the name of the Korean Association of Mathematics Education in the beginning. Park[17] said: "I don't remember the exact year when we changed the name from the Korean *Association* of Mathematics Education to the Korean *Society* of Mathematics Education; but it was about 1972. In fact, mathematics education was not regarded as a discipline in the beginning of the KSME in the 1960s. We used the term 'association' in the name, and later replaced it with 'society', as mathematics education had been accepted as a discipline in the 1970s" (p. 75). The name of the Korean Society of Mathematics Education was made by the first generation researchers in mathematics education, who made efforts to establish the KSME's identity. The KSME has steadily grown on the foundation of the senior scholars' efforts. Later in 2011, the KSME was registered as cooperation at the Seoul Metropolitan Office of Education. As of today, the KSME is one of major academic societies in mathematics and mathematics education both in amount of activities and in size. It has 1451 regular members and 104 institutional members so far.

The articles of the KSME states major responsibilities of the society: (a) To publish journals and books on mathematics education, (b) To host a national conference on mathematics education, and (c) to collect and display books, samples, and teaching/learning tools on mathematics education. To meet the first responsibility, the KSME currently publishes five periodical journals and the newsletters as well as publishes several books including *Math Club Activity Materials for Gifted Students, General Contents of The Mathematics Education (TME), Ideas of Analyzing Problem Solving, Teaching Materials for Underachieved Students in Mathematics,* and *List of the Articles in the KSME Journals.* Also, the KSME began to publish yearbooks from 2013 and the first yearbook's title is *Constructivism and mathematics education in Korea.* The KSME will continue to publish next issues every year.

For performing the second duty, the KSME has hosted national conferences on mathematics education every four months; one of three focuses on mathematics education for gifted students. This conference was expanded to the international conference in 2012. Many foreign mathematics education researchers have participated in the international session (spoken in English) to present their new findings, share different approaches, and communicate with Korean researchers. Table 9-1 shows the themes in the conferences, for example, from 2008 to 2013.

Table 9-1. National Mathematics Education Conferences hosted by of the KSME

Year & Month	Conference Theme
2008.2	A systemic plan for activating education for mathematically gifted and creative students
2008.3	Mathematics teaching by level: Reality and improvement
2008.11	Mathematics teaching by level: Reality and improvement (II)
2009.2	Mathematics education for training creative students
2009.3	Leading positive attitudes and motivation toward mathematics
2009.10	Increasing creativity in mathematics education
2010.2	Creative inquiries of mathematically gifted students
2010.4	*Joint conferences of mathematics related societies*
2010.10	Creatively realizing mathematics curriculum
2011.4	Vitalization of public education: School mathematics and teacher professionalism
2011.8	Developing creativity and personality of mathematically gifted students
2011.11	Mathematics education: Present status and prospect
2012.4	Developing professionalism of mathematics teachers
2012.8	*Not holding due to ICME 12*
2012.11	A new taking-off of mathematics
2013.4	A new horizon of mathematics education
2013.8	Research on and education for mathematically gifted students
2013.11	Teaching in a mathematics classroom

The KSME has well performed in its first and second mission but not well in the third mission, collecting books, materials, tools for mathematics teaching. The society needs to make more effort in this third mission.

On the other hand, researchers have discussed the development plans for the KSME[2,6,13,17]. They claimed (a) to increase the membership and encourage research activities, (b) to strengthen the international positioning of the KSME, (c) to make journals in high quality, and (d) to publish various practical and research materials of mathematics education. Throughout these efforts, the KSME will keep to perform a role as the center for developing mathematics education in Korea. In particular, as an effort for (b), in 2012, the KSME (with the KSESM) hosted the Twelfth International Congress on Mathematics Education (ICME 12) and hosted (with the Korean Society for History of Mathematics) the History and Pedagogy of Mathematics 2012 (HPM), an Satellite Meeting of ICME-12. These two international conferences were very successful.

3   Journals of the KSME and Math Education in Korea

3.1 KSME Journals

One of key roles for the KSME in regards to mathematics education in Korea is to publish journals through which the KSME encourages researchers to conduct research on mathematics education, strengthens researchers' potentiality, and disseminating research results. Currently, the KSME publishes five periodical journals. The KSME published the first issue of *The Mathematical Education* (*TME*) in March 1963, the first academic journal on mathematics education in Korea; most recently, Vol. 52 no. 4 (total no. 143) of *TME* was published in November, 2013. Figure 9-1 shows the cover of the first issue of the *TME*.

The first president of the KSME, Hansik Park[14] stated in the first issue of the Journal of the KSME that: "We will go forward by doing research and study, and for the moment we will investigate and review educational innovations used in developed countries. We have our own circumstance and foundation under which we should and can find an efficient way to mathematics education fitting better to us. We, mathematics education researchers, have a mission of playing a role as parameters in the KSME. I hope this effort will make a difference in mathematics education of Korea" (p. 1). Dr. Park claimed that we should not just follow Western forms of mathematics education but develop our own mathematics education based on our environments and our societal demands. He positioned the KSME and the journal of the KSME as a medium in which mathematics education researchers can communicate with each other. *TME* was published in three issues (March, July, November) a year for several years. Publication was reduced to two issues and sometimes even one issue a year in the late 1960s. Since 1975, two issues a year, and in 2003 the number had risen to four issues a year, periodically.

Today, the KSME publishes four additional journals. The *Pure and Applied Mathematics* (*PAM*) publishes research papers in all areas of mathematics; first issue in 1994, now four issues a year. The *Education of Primary School Mathematics* (*EPSM*) publishes research articles on elementary (from grade 1 to grade 6) mathematics education; three issues a year. The *Research in Mathematical Education* (RME) is the international journal published by the KSME; first issue 1997, four issues a year. The *RME* is the only academic journal written in only English in Korea. The *Communications of Mathematical Education* (CME) was first issued in 1990 and now four issues a year. The *CME* was like the proceedings of the KSME conferences in the 1990s, but it is now a peer-reviewed journal.

韓國數學教育會誌

数學教育

1 9 6 3. 3.

Volume 1. Number 1

THE
MATHEMATICAL EDUCATION

JOURNAL OF KOREA SOCIETY OF MATHEMATICAL EDUCATION

Published by the Association
(c/o College of Education, Seoul Nati. University, Seoul, Korea)

韓 國 數 學 教 育 會 發 行

Figure 9-1. Cover Page of the First Issue of TME

3.2 Articles Published in The Mathematical Education

Research history of mathematics education in Korea is alive in the articles in *TME*. Let us evaluate the articles published in *TME* from 1963. There are some differences between the articles in the 1960s and in the 1970s. Figure 9-1 presents the titles of the articles published in the first issue and Table 9-2 shows the titles of the articles published in Vol. 1, issue 2 (1963) and in Vol. 13, issue 3 (1975).

Table 9-2. Some Articles Published in *The Mathematical Education* Vol. 1 and Vol. 13

Vol. 1, no. 2 (1963)	Vol. 13, no 3 (1975)
On the evaluation of learning arithmetic	Mathematical education discussed at ICMI
	Points of teaching trigonometry
An instrument of solid geometry	On Young's inequality
Some problems in the high school mathematics	A Study of modernization of mathematical education
The international movement on attempts to modernize mathematics teaching	Note on the double centralizer of a module
	A note on metrization of topological spaces
Resolutions and suggestions from the ICME-2	On tightness of product space
	Rational extensions of modules and D rings
The mathematical terminology	On matroids

TME Vol.1, no. 1 and no. 2 included articles on just mathematics education, while Vol. 13, no. 3 included three articles on mathematics education and six articles on mathematics. In the 1970s up until the middle 1990s, *TME* had included research articles on mathematics education as well as on mathematics. In the 1980's, more than half of the articles in *TME* were on mathematics (see Table 9-3). In 1994, the KSME published a new journal, the *PAM*, for research articles in all areas of mathematics. Since then, mathematics education articles have been published in *TME* and mathematics articles in the *PAM*.

Table 9-3. Number of Articles Published in *TME* from 1977 to 1983

Year	Research area	
	Mathematics education	Mathematics
1977	7	12
1978	6	20
1979	2	5
1980	5	15
1981	6	19
1982	7	24
1983	6	40

Lee[12] explained why in *TME* the number of mathematics articles is more than that of mathematics education articles: "A university requires research activities in mathematics rather than in mathematics education when hiring and promoting the mathematics faculty members (in charge of mathematics education). So, professors did research and wrote papers in mathematics rather than in mathematics education" (p. 250). This assertion sounds correct in that there rarely were college faculties specializing in mathematics education in the 1970s and the 1980s, and researchers writing the articles published in *TME* specialized in a mathematical area. In addition, lack of research basis and professional researchers in mathematics education are other factors to publishing mathematics articles in the *TME*. Korean researchers have been educated at last in the middle of 1980s as doctoral programs in mathematics education opened in Seoul National University and the Korea National University of Education. In brief, the KSME had no choice but to publish mathematics articles in *TME* in the 1970s until the early 1990s due to lack of researchers, research potentials, and academic journals for mathematics education research. Currently, *TME* has had just mathematics education articles since 1995; most colleges of education across the nation open doctoral programs in mathematics education through which new researchers have been trained. They have actively presented research studies in the KSME and other conferences; the articles have been published in national and international academic journals.

By examining the articles published in *TME*, the articles in the 1960s are mostly documentations of empirical knowledge gained from practices and introduction to international trends (including the New Mathematics). So, we may conclude that the 1960s can be seen as a time period of cumulating empirical knowledge before theoretical conceptualization, and the same tendency would be taken until the middle 1980s. See the articles listed below from 1963 to 1966. The volume and the number of the articles were said only in the end of the last article of the issue. See Figure 9-1 for Vol. 1, no. 1; see Table 9-2 for Vol. 1, no. 2.

- A study and teaching improvement of the child's discriminative ability of the geometrical figures
- A study on learning habit of mathematical education
- An effective method of introducing differential calculus (Vol. 1, No. 3)
- A study on the scale of notion
- The opportunity and arrangement on teaching of the multiplication tables
- The experimental researches on teaching of quadratic equations (Vol. 2, No. 1)
- Case study of linear programs and branching programs on mathematics
- The teaching method of arithmetic at the dull class (Vol. 2, No. 2)
- On the teaching method of mathematics for the rural student
- The influence of fundamental abilities in learning mathematics (Vol. 2, No. 3)
- A study on the counting ability of the first grade pupils in their beginning months
- A study of the cases of computation errors and wrong answers in high school mathematics (Vol. 2, No. 4)
- The effect of the intuitional instruction for basic formation of the thinking construction in calculation I
- Some points in teaching probability
- Cross puzzle using sings U and ∩
- New thinking in mathematical education (Vol. 3, No. 1)
- The effect of intuitional instruction for basic formation of the thinking construction in calculation II
- The limit concept in the SMSG revised sample textbooks
- The studies on history of mathematical education in the 20th century (Vol. 3, No. 2)
- What is new method in teaching slow learners in middle school?
- Comments on UNESCO "The Teaching of Mathematics at the Secondary Levels" Preliminary Edition, June 1965
- Systematic analysis of sets for the modernization of mathematics education (Vol. 3, No. 3)
- A study on instruction of non-Euclidean geometry for mathematics education
- A research for teaching systematic probability introduced by sets (Vol. 4, No. 1)
- A study of teaching position of digits and writing of numbers
- A comparative study of program instructions with and without illustrations

- Congruency of triangles by AAS
- Problems on prime numbers (Vol. 4, No.2)
- Mathematics teaching in Afghanistan (Vol. 4, No.3)
- The expression of ratio and its three uses
- A study for development of calculating ability in arithmetic
- A study on the effectiveness of programmed learning in mathematics I
- The problems of modernization of mathematical education (Vol. 5, No. 1)
- A present status and teaching method of addition and subtraction ability at an outskirt school
- A study on the method of finding a cubed root
- A study on the effectiveness of programmed learning in mathematics II (Vol. 5, No. 2)

After the middle of 1980s, young scholars earned doctoral degrees in mathematics education in other countries like the U. S. have returned to Korea. Their genuine works were introduced in *TME*. Also, more scholars who have completed doctoral study in the Country (Korea) have joined in publishing research articles in *TME*. Those articles were different from the articles published before in both quality and quantity. See the list of articles in *TME* below from Vol. 33, no. 1 (1994, March) to Vol. 34, no. 1 (1995, June). The themes of these articles vary: mathematics education evaluation, students' individual differences, van Hiele theory of geometry learning, metacognition and mathematics education, technology in mathematics education, mathematics gifted student education, mathematical modeling, and comparison of mathematics textbooks between different counties. These articles were written under the theoretical knowledge basis, while the articles in the 1960s showed empirical knowledge:

- A study on the transition of high-school statistical education
- Mathematical idea of River Ages of the East and Sea Ages of the West
- Direction of a branch of mathematics in university entrance examination
- Evaluation for middle-school mathematics by the item response theory: Comparative study of difference between the school groups
- Measurement of the geometrical development level of university students by van Hiele theory
- A study on the differences in mathematical ability between male and female in senior high school (Vol. 33, No. 1)
- Designing a program for mathematically gifted student
- Mathematics resources on the internet
- Study on evaluation of mathematical ability when some data of criterion referenced evaluation model are missing
- Development of metaproblems for the instruction and evaluation of metacognitive problem-solving abilities
- A study on the current state of evaluation in secondary mathematics education

- A suggestion for improvement of mathematics education for engineering major college students
- Mathematics education through LOGO: Contents and methods
- New directions in the evaluation of mathematics
- A study of the effects of student's van Hiele level in teaching geometry introduced by the history of mathematics and quizzes
- A study of metacognition on the exploration stage of mathematical problem solving procedure for grade eight gifted students
- A study on geometric thinking levels of junior high school students according to van Hiele's theory
- The application of spreadsheet in mathematics education
- A study on the procedure and criterion for identifying the mathematical gifted students
- A study on the effective teaching methods for the 8th graders with mathematical anxiety (Vol. 33, No. 2)
- The reconceptualization of teacher education program in mathematics
- Computer aided mathematical proof in finding the inverse matrix of a special matrix
- The exploration of mathematics learning model on the basis of cognitive development
- The development of educational S/W for subset
- Some note on the paradox of Zenon
- The practical study of open-end approach in mathematics education
- The effect of explorative instruction using mathematical modeling
- Research on the use of calculator in elementary mathematics education
- A comparative study of mathematics textbooks of Korea and Russia: Laying stress on the geometry
- A comparative study of mathematics textbooks of Korea and Russia: Laying stress on the algebra
- The development of education S/W ESPDE for numerical solution of time-dependent patial differential equation (Vol. 34, No.1)

3.3 *TME* and Mathematics Curriculum

Korea has revised the mathematics curriculum several times since the Syllabus in 1946. The mathematics curriculum has not only reflected the results from mathematics education researches but also promotes research in mathematics education. Let us see how the curriculum revisions affect the research on mathematics education through *TME* articles. In this section, we focus on the sixth mathematics curriculum (MC-6) revised in 1992, which had a huge impact on mathematics education, and the seventh mathematics curriculum (MC-7) revised in 1997.

It is noted that the MC-6 presented the use of calculators and computers in teaching mathematics by stating that "a calculator or a computer in mathematics teaching *can be used* for complex calculation and for increasing the ability of solving problems" (Department of

Education[4], p. 239). Furthermore, the MC-7 suggested a more positive opinion on the use of technology in mathematics classrooms by stating that "except in the case of computational purpose, in order to perform complex computation, understand mathematical concepts, principles, and rules, and to increase problem solving ability, mathematics teaching *uses* technological tools such as a calculator, a computer, and other educational software along with other tools" (Department of Education, Science, and Technology[5], p. 46).

Let us see the mathematics education research articles published in the *TME* in the MC-6 and MC-7 time period. The listed articles below show all *TME* articles describing the use of calculators or computers before the MC-6. There were only six among which three articles were on CAI courseware, two articles computers and mathematics learning, and one article computers and the mathematics curriculum:

- On development of the computer-related mathematics learning in high school (1987.12)
- On the development of microcomputer-assisted mathematics teaching / learning method (1988.12)
- Role of CAI teaching mathematics and problem solving (1989.6)
- Influence of the computers on curriculum of mathematics (1990.12)
- A study on the CAI courseware for teaching arithmetical operations (1991.12)
- The problems and solution of courseware used in mathematical education (1992.6)

Even after the MC-6, research studies related to the use of calculators and computers were rarely conducted. In reality, schools did not have personal computers for students in those days. Considering the school environments of no computer classroom, the shortage of teachers implementing computer-assisted innovations, and no preceding research studies on use of computers, the MC-6 was not prepared by the needs of schools and research but it just followed the international trends. In particular, the Curriculum and Evaluation Standards for School Mathematics[19], translated into Korean in the August of 1992, stressed the use of calculators and computers in mathematics education. Having influenced many Korean educators and researchers, the use of calculators/computers was introduced in the MC-6. All *TME* articles describing the use of calculators or computers from 1993 to 1997 when the MC-6 was implemented are listed below:

- Courseware for Mathematical Education Using Artificial Intelligence (1993.6)
- Trends in computer literacy education (1993.12)
- Mathematics Resources on the Internet (1994.12)
- Mathematics Education through LOGO: Contents and Methods (1994.12)
- The application of Spreadsheet in Mathematics Education (1994.12)
- Research on the Use of Calculator in Elementary Mathematics Education (1995.6)
- Use of a computer in mathematics education: Teaching Calculus using the Mathematica (1996. 6)
- A study on function teaching using a graphic calculator (1997.6)
- A study on the effect on learning using CAI: An effective method to teach Pythagorean Theorem (1997. 6)
- Statistics education using motion graphics of XLISP-STAT (1997.12)

The MC-7 revised in 1997 has the key characteristics of the level-differentiated mathematics education. The fundamental idea in the MC-7 is to "construct the curriculum in which a student's learning ability/skills and contents reciprocally work together so that the curriculum can provide mathematical contents adjusted to the level of learning. ... There were various types: the stepwise form, the enrichment/supplement form, and the optional subject form" (Hwang, Na, Choe, Park, Lim, Seo[7], p. 77).

Table 9-4. Articles on the Level-Differentiated Education Published in *TME* from 1990 to 2002

Year Month	Title
1997.12	A plan for step-type differentiated teaching of middle school mathematics
1998.11	A new direction of assessment in the curriculum for different student's levels
2000.5	The effect of self-directed learning by self-selecting of the level tasks for the students' own level on achievement in mathematics
2001.11	A study on formation & application of step-wise level curriculum of mathematics

As seen in Table 9-4, there were no articles about the level-differentiated mathematics education until 1997, the year of MC-7. Even after the MC-7, such articles were just a few. The level-differentiated mathematics education was not prepared from our own demands but injected from the outside. Thus, many problems arose when implementing the MC-7. The Korea Institute for Curriculum and Evaluation (KICE) reported the problems in dividing a class for enrichment and supplement, separating classes for different levels, forming level-differentiated class, textbooks for different levels, evaluation, etc. For example, KICE (2004) stated that:

> The current mathematics textbooks include the normal course and the enrichment course together. Students and parents genuinely wish to learn all the contents within the textbooks. This has caused an exceeded normal amount of learning. A textbook does not provide content for supplementary course. Mathematics teachers develop their own materials so the level and the quality are not consistent. It also requires teachers' extra work; therefore they have complaints (p. 35).

Such a problem occurred first due to the lack of research studies and preparedness. The level-differentiated education is still emphasized in the mathematics curriculum revised in 2009 so it is now to research useful plans and implementing methods.

4   Conclusion

The KSME was established in 1962 with the goal of collecting and analyzing information on the trends of international mathematics education, evaluating the mathematics education of Korea, and then suggesting reforms of mathematics education in Korea. The KSME focuses on publishing journals and books as well as hosting national and international conferences. Currently, the KSME periodically publishes five journals, a newsletter, and hosts conferences three times a year. The KSME publishes *TME*, the first academic journal of mathematics education in Korea; the first issue was published in March, 1963; there are a total of 143 issues as of November, 2013. The articles published in *TME* are mainly about mathematics instructions, mathematics teaching materials and the use of technological education tools, and mathematics education psychology. While the articles in the 1960s delivered empirical knowledge, after the late 1980s, articles have been written with a theoretical approach.

Articles published in *TME* reflected the MC-6 and the MC-7. The key characteristic of the 'MC-6 Revised' in 1992 was the use of calculators and computers, and the 'MC-7 Revised' in 1997 stressed differentiated education. Analyzing *TME* articles about the use of calculators/computers or differentiated education revealed that the MC-6 and the MC-7 were not solely products of Korea's own research results but was largely influenced by international trends.

This article is just a start to find characteristics in development of mathematics education in Korea putting focus on the role of the Korean Society of Mathematical Education and its journal The Mathematical Education for the past half century. Today, Korea is regarded as one of scientific counties in Asia and even in the world, which have research potentials in mathematics education. In reality and in name, Korea becomes rapidly a hub of East-Asian region. These productive results no doubt come from our senior scholars true efforts. This article is a summary on how those efforts contributed to the development of mathematics education in Korea in kind of a historical view. Hopefully, different views will uncover other characteristics in the progress of mathematics education. These reflective and evaluative works can help guide a way on which mathematics education of Korea will go forward in the new era.

References

1. Choe, Y. H. (1969). A study on history of mathematics education of Korea. *The Mathematical Education, 8*(1), 18-25. [in Korean]
2. Choe, Y. H. (1996). The way Korea Society of Mathematical Education should take. *The Newsletter of Korea Society of Mathematical Education, 12*(1), 12-14. [in Korean]
3. Department of Culture and Education (1963). *High school curriculum commentary.* Seoul: Author. [in Korean]
4. Department of Education (2000). *A standard of school mathematics curriculum.* Seoul: Author. [in Korean]
5. Department of Education, Science, Technology (2011). *Mathematics curriculum.* Seoul: Author. [in Korean]
6. Han, I (2010). On the trends of the Korean Society of Mathematical Education leading mathematics education. *Proceeding of the 45th Korean National Meeting of Mathematics Education*, (pp. 87-89). [in Korean]
7. Hwang, H. J., Na, G.S., Choi, S.H., Park, K.M., Lim, J.H., Seo, D.Y. (2012). *Mathematics education: New approaches.* Seoul: Mooneumsa. [in Korean]
8. Kilpatrick, J. A history of research in mathematics education. In D. A. Grouws (Ed.), *Handbook of research in mathematics teaching and learning* (pp. 1-38). NY: Macmillan Publishing Company.
9. Kim, Y.U., Kim, Y.K. (1982). *History of Korean mathematics.* Seoul: Seolhwadang. [in Korean]

10. Korea Institute for Curriculum and Evaluation (2004). *Mathematics curriculum: Reality and reform.* Seoul: Author. [in Korean].

11. Lee, J. K. (2004). *Mathematics curriculum of Korea.* Seoul: Kyungmoonsa. [in Korean]

12. Lee, K. S. (2003). A Classification and analysis of the articles in 'The Mathematical Education' from issue 1 to issue 99. *The Mathematical Education, 42*(2), 247-258. [in Korean]

13. Lee, K.S. (2012). Overview of the Korea Society of Mathematical Education 50 years. *Proceedings of the KSME 2012 Spring Conference on Mathematics Education,* (p. 1). [in Korean]

14. Park, H.S. (1963). Let us find the king's road. *The Mathematical Education, 1*(1), 1. [in Korean]

15. Park, H.S. (1982). *History of mathematics education.* Seoul: Gyohaksa. [in Korean]

16. Park, H.S. (1993). Passage of the Korea Society of Mathematical Education for past 30 years. *The Mathematical Education, 32*(3), 1-6. [in Korean]

17. Park, H.S. (2004). Participating in the ICME. *Communications of Mathematical Education, 18*(2), 1-8. [in Korean]

18. Park, H.S. (2012). Necessity of an academic society. *Proceedings of the KSME 2012 Spring Conference on Mathematics Education,* (pp. 11-12). [in Korean]

19. NCTM (1989). *Curriculum and evaluation standards for school mathematics.* Reston, VA: Author.

CHAPTER 10

CURRICULUM REFORM AND RESEARCH TRENDS IN EARLY CHILDHOOD MATHEMATICS EDUCATION IN KOREA

Ji-Eun Lee

Department of Teacher Development and Educational Studies
Oakland University
2200 N. Squirrel Road, Rochester, MI USA
E-mail: lee2345@oakland.edu

This chapter introduces contemporary curriculum reform and research trends in early childhood mathematics education in Korea. This chapter consists of two major topics. First, it examines the Nuri curriculum, a recently developed and implemented national-level early childhood curriculum. The examination of the principles and contents of the Nuri curriculum intends to elicit a place for early childhood mathematics education in the Korean context. Second, this chapter examines research trends in early childhood mathematics education in the recent four years (2010–2013). It is expected that this review will provide some insights into the agenda for future research to benefit from recent curriculum changes and to offer suggestions for improving the quality of the early childhood mathematics education curriculum.

1 Introduction

Teaching mathematics has long been a vital element in early childhood education[3]. Regardless of some persistent misconceptions and internally or externally challenging factors[6,19], the importance of early childhood mathematics education in the lives of young children has been recognized globally[9,10,11]. Correspondingly, it is generally supported that teaching mathematics as early as possible is beneficial in many ways[13]. Early childhood mathematics education typically focuses on a few essential mathematical concepts and processes in an integrated way using everyday activities, rather than being taught as an isolated subject[4,5,7,12]. The impact of early childhood mathematics education has been reported in many longitudinal studies. Research generally supports that mathematical understanding and competency in the early years can be a reliable predictor of later mathematical competence[1,2].

In Korea, there have been continuous efforts to effectively integrate mathematics in early childhood education as evidenced in multiple curriculum reforms and educators' research over the years. This chapter intends to introduce contemporary research trends in early childhood mathematics education in Korea in two ways. First, it examines the overall structure, principles, and content of the most recent early childhood education curriculum focusing on the area of mathematical exploration and inquiry. When substantial changes occur in the education system, all aspects of teaching and learning will require considerable reexamination[16]. The advent of the Nuri curriculum, a recently developed and implemented national-level early childhood curriculum in Korea, naturally calls for the need to examine the planned and enacted curriculum. The examination of guiding principles, contents, and suggested activities in the Nuri curriculum would be a timely effort and helpful in understanding the place and role of mathematics education in early childhood education in Korea.

Second, this chapter reviews recent research trends in early childhood mathematics education in Korea based on articles published over the last four years (2010–2013). This review is expected to provide the background for understanding what has been studied in the Korean context. At the same time, it provides a basis for looking forward into the next step. In particular, this review provides some insights into potential agendas for future research to benefit from the recent curriculum changes and to offer suggestions for improving the quality of early childhood mathematics education curriculum.

2   Early Childhood Education and Care in Korea

2.1 *A Brief Overview of Recent Reform in Early Childhood Education*

Early Childhood Education and Care in Korea began over 100 years ago. Programs for young children have been developed and governed by two distinctive administrative bodies for a long time. One type is the kindergarten program for children aged three to five before elementary education under the administration of the Ministry of Education, Science, and Technology. Another is a program offered at childcare facilities under the administration of the Ministry of Health and Welfare and Family Affairs[23]. Originally, the former more emphasized the function of young children's *education*, whereas the latter put more emphasis on *care* of children. However, the distinction between these two types of early childhood education programs has become obscured in practice because kindergartens began to operate full-day programs by including the *caring* function with the Early Childhood Education Promotion Act since 1992, and child care centers have tried to strengthen their educational role[24,28].

As a collaborative effort to improve Early Childhood Education and Care, a common national-level early childhood curriculum has been developed, "which universally applies to all children aged between 3-5 in both kindergartens and childcare centers through integrating two separate curriculums into one, the Nuri curriculum"[8]. The literal meaning of "Nuri" is "world." This name signifies "a wish for all children to lead their happy lives and fulfill all of their hopes and dreams"[8] and its goal of promoting holistic development of young children. The Nuri curriculum for 5 year-old children was implemented in 2012. Later, this curriculum was extended to younger children as well. Since March 2013, all children ages 3-5 attending either type of early childhood education or care institution have been using the Nuri curriculum and received subsidies. In a practical sense, this policy yielded an equivalent effect of expanding the years of compulsory education in Korea.

2.2 *The Nuri Curriculum: A Big Picture*

Prior to the arrival of the Nuri curriculum, kindergartens and childcare centers each operated under a different set of curriculums and guidelines. The kindergarten curriculum has gone through multiple revisions since it was first established in 1969. Although the emphasis has been refined in each revision, typically the kindergarten curriculum has highlighted the basic abilities and attitudes needed in everyday life. For example, the following five domains were included in the sixth revision implemented in 2000: physical health in daily life, social relationship in daily life, expression in daily life, language in daily life, and inquiry in daily life. Childcare program services typically include care, education, nutrition, health, safety, services for parents, and exchanges with communities. Childcare programs for children ages 3-5 need to consider the following basic components when establishing and implementing instructional plans: basic abilities in daily life, physical activity, social relationships, communication, experience in art, and nature and discovery[24]. The unification of kindergarten curriculum and childcare center programs as reflected in the Nuri curriculum consists of five areas (Ministry of Education, Science and Technology [MOEST] & Ministry of Health and Welfare [MOHW][21]):

- Physical activity and health: Developing basic physical abilities and establishing healthy and safe routines.
- Communication: Learning how to communicate in daily life and developing good practices in language use.
- Experience in art: Developing interest in aesthetics, enjoying the arts and learning how to express oneself creatively.
- Social relationships: Developing self-respect and learning how to live with others.
- Nature and discovery: Exploring the world with curiosity and enhancing children's abilities to solve problems by applying math and science in daily life.

The following guidelines are suggested for the actual operation and implementation of the curriculum. For organization of programs, balanced and integrated coverage of five main areas, play-based activities, flexibility, and equitable and unbiased approaches are suggested. For the actual operation and implementation of the program, utilization of various indoor and outdoor activities, consideration of children's ability and levels of disabilities, educational opportunities for parents, collaboration with family and local communities, and re-training opportunities for teachers are addressed. For the actual instruction, play-based and child-centered teaching and learning, authentic learning in daily life, various interactions (e.g., teacher-child, child-child, child-environment), and integrated approaches are suggested (MOEST & MOHW[21], pp. 21-32).

2.3 *The Place of Mathematics in the Nuri Curriculum*

To present the place of early childhood mathematics in the Nuri curriculum, this section introduces the content categories in five main areas and specific content expectations in mathematics. Table 10-1 illustrates these five main areas and content categories as described in the Nuri curriculum (see MOEST & MOHW[21] for more detailed information).

Table 10-1. The Content of the Nuri Curriculum: Main Areas and Content Categories

Physical activity and health	Communication	Experience in art	Social relationships	Nature and discovery
• Physical self-awareness • Motor control and basic exercise • Participation in physical activities • Healthy living habits • Safety habits	• Listening • Speaking • Reading • Writing	• Discovering beauty • Expressing self through art • Appreciating arts	• Building self-esteem • Understanding others and self-awareness • Cherishing family • Living in harmony with neighbors • Stimulating interest in society	• Promoting inquisitive attitudes • ***Mathematical exploration and inquiry*** • Scientific exploration and inquiry

Note: Adapted from "Nuri curriculum: The first step toward the integration of the split systems of early childhood education and care in Korea. KICCE Policy Brief, 2," by M. Chang, 2013. Korea Institute of Child Care and Education.

As presented in Table 10-1, "mathematical exploration and inquiry" is one of 20 categories in the Nuri curriculum, part of the "Nature and Discovery" area. Table 10-2 provides more detailed content expectations by age in this category (see MOEST & MOHW[22] for more detailed information).

Table 10-2. Content Strands and Expectations in Mathematical Exploration and Inquiry

Strand 1. Understand Basic Concepts of Number and Operations

Age 3	Age 4	Age 5
Become interested in numbers in daily life.	Understand the meaning of numbers used in daily life.	
Compare the quantities of objects.	Understand 'equal to', 'more than', and 'less than' in comparison of the quantities of objects.	Understand the part-whole relationship in quantities of objects.
Count up to 5 concrete objects and become interested in quantity.	Count up to 10 concrete objects and identify the quantity.	Count up to 20 concrete objects and identify the quantity.
		Experience adding and subtracting with concrete objects.

Strand 2. Understand Basic Concepts of Space and Geometric Shapes

Understand in front of, behind, next to, above, below myself.	Represent location and direction using various methods.	
		View object from different orientations and compare the differences.
Become interested in the shape of object.	Recognize the characteristics of basic geometric shapes.	Understand the similarities and differences of basic geometric shapes.
	Construct various shapes using basic geometric shapes.	

Strand 3. Basic Measurement

Compare two objects by length.	Compare length, size, and weight in daily life.	Compare length, size, weight, and capacity in daily life and order them.
		Measure length, area, capacity, and weight using non-standard measuring units.

Strand 4. Understanding Patterns

Become interested in the repeated patterns in daily life.	Understand the repeated patterns in daily life.	Understand the repeated patterns in daily life and be able to extend them.
	Recognize repeated patterns and follow the rule.	Independently Create patterns.

Strand 5. Basic Data Collection and Analysis

	Collect needed information and materials.	
Matching same objects.	Sort objects based on one rule.	Sort objects based on one rule and re-sort them using a different rule.
		Create graphs using pictures, photos, symbols, or numbers.

As shown in Table 10-2, differentiated content expectations for three age groups consider the developmental level of each age group and are presented under uniform major strands. For example, similar content can be introduced in a varying level of complexity as shown in Table 10-3.

Table 10-3. An Example of Instructional Sequence by Age

Age 3	Age 4	Age 5
Prepare two sets of objects that are visibly different in quantity. Ask children to compare the quantities of two sets.	Prepare objects less than 10. Ask children to make the same quantity using other objects. Prepare two sets of objects that contain same amount of objects. Add 1-2 objects to one of the sets. Ask children to compare the quantities of two sets.	Decompose about 10 objects using various methods (e.g., 6 yellow marbles and 4 red marbles; 3 yellow marbles, 2 red marbles, and 5 green marbles). Compose groups of objects to make a certain number (e.g., making a 5 game).

Note: Adapted from "The Nuri curriculum for 3 – 5 years: Guidelines for teacher" (p.151), MOEST & MOHW, 2013b, Seoul, Korea.

As continuously reiterated in many official documents[21,22], these standards should be understood as a set of guidelines for designing actual instruction. Teachers are expected to flexibly construct their daily instruction based on children's needs and interests by taking an integrative approach[20]. Also, it should be noted that although more detailed content standards are being introduced in the area of mathematics, these are included under the main principles of early childhood education. These principles include child-centered, integrated, play-based, and developmentally appropriate approaches.

The development and implementation of the Nuri curriculum demonstrates recent research efforts to enhance the quality of early childhood education in Korea. At the same time, the new guidelines and standards are expected to generate new research agendas. In this regard, it would be meaningful to examine the recent research trends.

3 Trends in Early Childhood Mathematics Education Research in Korea: 2010–2013

This section surveys recent trends in early childhood mathematics education in Korea by examining the major topics, research methods, research participants, and implications. This overview is mainly based on articles published in Korea between 2010 and 2013. This period is particularly significant not only because these are the most recent years but also because they represents a transitional period to a new national curriculum. The following sub-sections briefly report on the results of analysis focusing on *what, how, and whom* these articles studied.

3.1 *Trends in Productivity and Outlet for Research*

For the purpose of this overview, 54 articles from 17 journals published in Korea between 2010 and 2013 were selected (12 articles in 2010, 11 articles in 2011, 9 articles in 2012, and 21 articles in 2013). Initially, the articles were searched via Google Scholar and collected through multiple databases (see Appendix for the list of journals). This overview includes only the studies specifically related to early childhood mathematics education. Forty-five out of 54 articles have been published in 10 journals specializing in early childhood education. The rest were published in journals specializing in mathematics education, general teacher education, general education, computer game, and home education.

The number of published articles notably increased in 2013. It is possible that the discussion on and preparation for the new national-level early childhood education curriculum might be the contributing factor to explain this increase. However, since this chapter only selected articles published during the last four years, it would not be appropriate to make a definite conclusion on the pattern of research productivity.

As reflected in the specialties of journals, most of the early childhood mathematics education research has been conducted within the early childhood education community. Only a few research studies examined issues related to early childhood mathematics education in different fields (e.g., technology, special education).

3.2 Trends in Research Topics

Based on the methodology of content analysis, the research topics related to early childhood mathematics education were classified into 10 categories along with sub-categories using the inductive content analysis method[15]. The final analysis framework consisted of the following research topic categories and sub-categories as shown in Table 10-4. Since many articles addressed more than one category, one article might be coded into multiple categories.

Table 10-4. Frequencies of Topics in Articles

Category/ Sub-category [total frequency]	Frequency by year			
	Y10	Y11	Y12	Y13
A. Student competency and motivation **[25]**				
A1. Knowledge and ability [15]	7	2	2	4
A2. Attitude and disposition [9]	3	2	2	2
A3. Creativity [1]	1			
B. Effects of teaching strategies **[24]**				
B1. Effects of integration [14]	9	3	1	1
B2. Discourse (e.g., student discussion, teacher questioning) [3]	1			2
B3. Representation [2]	1			1
B4. Cooperative problem solving/ Small group activities [2]	1			1
B5. Manipulatives [1]				1
B6. General [2]	2			
C. Teacher competency and motivation (pre & in-service teachers) **[18]**				
C1. Beliefs, efficacy, attitude, anxiety [12]	2	5	1	4
C2. Knowledge [6]	3	1	2	
D. Teacher training program development and effects **[7]**				
D1. In-service professional development [1]			1	
D2. Pre-service teacher education [6]	1	3	1	1
E. Home connection **[6]**				
E1. Parent perceptions [3]			2	1
E2. Home environment [3]			2	1
F. Math content **[5]**				
F1. Number and operations [2]				2
F2. Geometry [2]			1	1
F3. Measurement [1]	1			
G. Analysis of National curriculum **[4]**				

G1. Curricular alignment with elementary curriculum [1]				1
G2. Analysis of content and activities [3]		1	1	1
H. Instrument development/Assessment criteria development [4]				
H1. Large-scale [2]			1	1
H2. Performance assessment [2]		1	1	
I. Meta analysis [2]				
I1. Program effectiveness [1]			1	
I2. Effects of math activities [1]	1			
J. Development and effect of math programs [2]				2

As represented in Table 10-4, two most popular topic categories in the reviewed articles were "student competency and motivation" (category A) and "effects of teaching strategies" (category B). Since changes in the level of students' competency and motivation can be influenced by methods of teaching and learning, and instructional strategies are typically designed to show their impact on students' performance, many articles were categorized in both topics. "Teacher competency and motivation" (category C) was the next popular research topic. On the whole, the research on teachers, students, and the interaction between them in the instructional setting has continuously been investigated.

In the investigation of effects of teaching strategies (category B), the majority of research studies investigated the effectiveness of various integrated approaches used to teach mathematics (category B1). Some examples of integration included art, music, science, physical activities, play, technology, storybooks, and games. While integration was the dominant strategy examined in this category, other aspects of teaching effectiveness tend to appear in the most recent year, such as investigation of class discourse, grouping strategies, representation, and instructional materials.

Regarding the math content (category F), three content strands (number and operations, concepts of space and geometric shapes, and measurement) were addressed. Two other strands from the Nuri curriculum, "patterns" and "basic data collection and analysis", were not included.

Analysis of the national curriculum (category G) focused on two areas. The first area examined the continuity of the mathematical exploration and inquiry suggested in the Nuri curriculum and, the elementary level mathematics curriculum. The second area analyzed mathematics-related content and activities presented in the teachers' guidebooks and other curricular materials. These studies focused on identifying both positive aspects of the current curriculum as well as weaknesses or areas to be improved.

3.3 *Methods of Research and Research Subjects*

Overall, about 60% of the articles mainly employed a quantitative research approach, and about 15 % of the articles were mainly qualitative. About 25% were theoretical or content analysis of extant literature or official documents. This section briefly addresses a few characteristics of the research methods employed.

The most frequently used research method was experiment/quasi-experiment (15 articles). These studies provided specific instructional interventions over a period of time and verified their effectiveness by comparing pre- and post-tests. Out of 15 articles, only one study involved pre-service teachers in a university setting. The remaining 14 studies were conducted in early childhood education classrooms with children. The sample participant size in these studies ranged from 24 to 82, and the duration of intervention ranged from 6 to 12 weeks. These studies generally concluded that the interventions used were effective.

The next frequent research method was descriptive research, which collected data using questionnaires and statistically analyzed the participants' responses to describe, explain, validate, or explore the target issues (14 articles). Participants in this type of research included teachers, pre-service teachers, and parents. The range of respondents in these studies was 40-710, and the average was 289 respondents. These studies served as need assessments by providing baseline data or were used as the basis for future program development.

In contrast to the studies that employed a quantitative analysis approach, the participants in the qualitative research studies were mainly in-service teachers. Out of eight studies, except for one with pre-service teachers and one with children, all of them were conducted with in-service teachers using observations, interviews, and analysis of various artifacts.

4 Implications for Future Research

This chapter introduced mathematical content in the recent early childhood education curriculum and research trends as an effort to provide a snapshot of the contemporary status of early childhood mathematics education in Korea. Curriculum reform and research are interrelated. Curriculum reform is based on research, and, at the same time, the new curriculum provides a context for research and makes suggestions to refine curriculum. Considering this on-going cycle, it would be meaningful to discuss future research directions based on a review of current curriculum and research trends.

4.1 *Need for Diversification in Research Topics*

As discussed in Section 3.2, although many research topics have appeared in recent trends, there are only a few areas that the majority of studies focused on, including student competency and motivation, effectiveness of teaching strategies, and teacher competency and motivation. While these are important topics and should continue to be investigated, more diversified research topics are expected. For example, the Nuri curriculum supports differentiated instruction based on children's ability and level of disabilities[21], but this topic has not been intensively addressed in the articles reviewed.

The Nuri curriculum also suggests equitable and unbiased approaches, and this suggestion is in line with previous studies in a foreign context and other recommendations[18,27]. However, the topic of addressing the needs and concerns of diverse children and families in early childhood mathematics education has not been found in this review.

While the Nuri curriculum suggests five strands in mathematical exploration and inquiry, an investigation of two strands (patterns and basic data collection/analysis) has not been found. It would be desirable to consider less-researched content areas in future research.

Other topics, such as developing and evaluating teacher education programs or investigating the role of families and communities to enhance early childhood mathematics education, are emerging but under-researched. There should be continued efforts to further examine these topics.

4.2 *Need for Diversification in Research Design and Methods*

As discussed in Section 3.3, many research studies employed a few dominant research designs and methods, such as experimental research and descriptive research based on questionnaire responses. Undoubtedly, these approaches have yielded valuable information to enhance early childhood mathematics education. However, a variety of research methods that compensate for the limitations of typical experimental or descriptive research should be employed. Noting the continued emphasis on the integrated approach in early childhood mathematics education[21], it would be desirable to obtain authentic and substantial data by utilizing diversified research designs. For example, several studies have pointed out the possible inconsistency between teachers' beliefs and actual actions[17,25,26]. In other words, what participants stated in a self-reported questionnaire may or may not be an accurate portrayal of their actions in the real setting. By diversifying research methods, it is expected that a more accurate picture of needs can be identified and more valid research will be produced.

4.3 *Need for Identifying Long-Term Trajectories*

In the discussion of formative assessment in mathematics education, Ginsburg[14] addressed the importance of teachers' understanding of three types of developmental trajectory. One is "normative information" concerning children's performance (e.g., knowledge of what first graders can normally do). Another is the understanding of the "cognitive trajectory" (e.g., progression in counting and number sense). The last one is the teacher's understanding of the "trajectory of mathematical ideas" (e.g., understanding ideas underlying the child's performance). This claim implies that children's mathematical performance should be understood holistically, considering both vertical progression and horizontal correlation across the curriculum. In this regard, the analysis of national early childhood curriculum (category G in Table 10-4) seems to be a timely effort. In fact, the examination of the mathematical contents and practices in the Nuri curriculum is currently underway through various community forums. It is expected that more research studies will be produced about this topic in the near future and will ultimately contribute to refining the existing curriculum.

In order to fully understand and confirm children's developmental trajectories, it would be helpful to get data from longitudinal studies. Most of the research studies reviewed in this chapter are based on one-shot questionnaires or one-time interventions. It is expected that long-term research can provide more rich and integrated data for future curriculum development and evidence-based practices.

4.4 *Need for Collaborative Research Efforts*

As indicated in the guidelines for program operation and implementation, the Nuri curriculum highlights collaboration with family and local communities and emphasizes the balanced and integrated coverage of five main areas[21]. This implies that there is a need for collaborative research efforts. However, interdisciplinary collaborative research is not visible. Most of the research topics are generally discussed within the early childhood education community as evidenced in the outlet of research studies. While this can be the most natural phenomenon (i.e., early childhood educators research early childhood mathematics), more active collaboration with scholars in other fields, such as special education, instructional technology, and mathematics education, would produce much richer information. It is expected that collaborative research efforts can create the synergy needed to provide the better opportunities for the children's success.

5 Looking Ahead

This chapter reviewed the recent curriculum reform and research trends in early childhood mathematics education in Korea. In order to benefit from a curriculum reform, it is necessary to have continuous communication and discussion among teachers, parents, children, educational researchers, and policy-makers. Much public discussion is currently underway over the implementation of the Nuri curriculum. Early childhood mathematics education should be understood as part of a larger curricular, social, and educational context. As this occurs, new research agendas will be generated and their findings will contribute to revisiting or refining the current curriculum.

References

1. Aunio, P., & Niemivirta, M. (2010). Predicting children's mathematical performance in grade one by early numeracy. *Learning and Individual Difference, 20*(5), 427- 435.

2. Aunola, K., Leskinen, E., Lerkkanene, M., & Nurmi, J. (2004). Developmental dynamics of math performance from preschool to grade 2. *Journal of Educational Psychology, 96*(4), 699-713.

3. Balfanz, R. (1999). Why do we teach children so little mathematics? Some historical considerations. In J. V. Coley (Ed.), *Mathematics in the early years* (pp. 3-10). Reston, VA: National Council of Teachers of Mathematics.

4. Björklund, C. (2008). Toddlers' opportunities to learn mathematics. *International Journal of Early Childhood, 40*(1), 81-95.

5. Björklund, C. (2010). Broadening the horizon: Toddlers' strategies for learning mathematics. *International Journal of Early Years Education, 18*(1), 71-84.

6. Bredekamp, S. (2004). Standards for preschool and kindergarten mathematics education. In D. H. Clements and J. Sarama (Eds.), *Engaging Young Children in Mathematics: Standards for Early Childhood Mathematics Education* (pp. 77–82). Mahwah, NJ: Lawrence Erlbaum Associates.

7. Brenneman, K., Stevenson-Boyd, J., & Frede, E. C. (2009). Math and Science in Preschool: Policies and Practice. *Policy Brief, 19.* New Brunswick, NJ: National Institute for Early Education Research.

8. Chang, M. (2013). Nuri curriculum: The first step toward the integration of the split systems of early childhood education and care in Korea. *KICCE Policy Brief, 2.* Korea Institute of Child Care and Education.

9. Clements, D.H., & Sarama, J. (2007). Early childhood mathematics learning. In F.K. Lester, Jr.(Ed.), *Second handbook of research on mathematics teaching and learning* (pp. 461-555). New York: Information Age Publishing.

10. Fox, J. (2007). International perspectives on early years mathematics. In J. Watson & K. Beswick (Eds.). *Mathematics: Essential research, essential practice.*

Proceedings from the 30th annual conference of mathematics education research group of Australasia Volume 2 (pp. 865-869). Tasmania, Australia.

11. Ginsburg, H. P., Cannon, J., Eisenband, J., & Pappas, S. (2006). Mathematical thinking and learning. In K. McCartney & D. Phillips (Eds.), *Blackwell handbook on early childhood development* (pp. 208–230). Malden, MA: Basil Blackwell.

12. Ginsburg, H. P., & Amit, M. & (2008). What is teaching mathematics to young children?: A theoretical perspective and case study. *Journal of Applied Developmental Psychology, 29*(4), 274-285.

13. Ginsburg, H. P., Lee, J., & Boyd, J. S. (2008). Mathematics education for young children: What it is and how to promote it. *Social Policy Report of the Society for Research in Child Development, 22*(1), pp. 3-23.

14. Ginsburg, H. P. (2009). The challenge of formative assessment in mathematics education: Children's minds, teachers' minds. *Human Development, 52*(2), 109-128.

15. Grbich, C. (2007). *Qualitative data analysis: An introduction.* Thousand Oaks, CA: Sage.

16. Kim, J., Colen, Y. S., & Colen, J. (2013). Reform based instruction in Korea: Looking over its promises to discover its sussesses. In J, Kim, I. Han, M. Park, & J. Lee (Eds.), *Mathematics education in Korea: Curricular and teaching and learning practices* (pp. 104-129). Singapore: World Scientific Publishing Co.

17. Leatham, K. R. (2006). Viewing mathematics teachers' beliefs as sensible systems. *Journal of Mathematics Teacher Education, 9*(1), 91-102.

18. Lee, J., & Ginsburg, H. P. (2007) Preschool teachers' beliefs about appropriate early literacy and mathematics education for low- and middle-socioeconomic status children. *Early Education and Development, 18*(1), 111-143.

19. Lee, J., & Ginsburg, H. P. (2009). Early childhood teachers' misconceptions about mathematics education for young children in the United States. *Australasian Journal of Early Childhood, 34*(4), 37-45.

20. Lee, K. (2000). Early Childhood Education Curriculum (2nd ed.). Seoul: Kyomunsa. [in Korean]

21. Ministry of Education, Science and Technology & Ministry of Health and Welfare (2013a). The Nuri curriculum for 3 – 5 years. Seoul, Korea: Author. [in Korean]

22. Ministry of Education, Science and Technology & Ministry of Health and Welfare (2013b). The Nuri curriculum for 3 – 5 years: Guidelines for teachers: Seoul, Korea: Author. [in Korean]

23. Na, J., & Moon, M. (2003a). Integrating policies and systems for early childhood education and care: The case of the Republic of Korea. *UNESCO Early Childhood and Family Policy Series, No. 7.* Paris: UNESCO.

24. Na, J. & Moon, M. (2003b). *Early childhood education and care policies in the Republic of Korea: Background report.* Seoul, Korea: Korean Educational Development Institute & Korean Ministry of Education and Human Resources Development.

25. Raymond, A. M. (1997). Inconsistency between a beginning elementary school teacher's mathematics beliefs and teaching practice. *Journal for Research in Mathematics Education, 28*(5), 550-576.
26. Skott, J. (2001). The emerging practices of a novice teacher: The roles of his school mathematics images. *Journal of Mathematics Teacher Education, 4*(1), 3–28.
27. Taguma, M., Litjens, I., Kim, J. H., & Makowiecki, K. (2012). *Quality matters in early childhood education and care: Korea.* OECD Publishing. Retrieved from http://dx.doi.org/10.1787/9789264175648-en.
28. Yun, E. (2009). *Places for educating and caring for young children in Korea: Where are our children edu-cared?* Child Research Net: ECEC around the world. Retrieved from http://www.childresearch.net/projects/ecec/2009_06.html.

Appendix: List of Journals

Journal Title	Number of articles reviewed
Journal of Early Childhood Education	13
Early Childhood Education Research & Review	12
The Journal of Korea Open Association for Early Childhood Education	8
Korea Journal of Child Care And Education	3
The Journal of Korea Early Childhood Education	3
Korean Journal of Human Ecology	3
The Korean Journal of Child Education	2
Journal of Korean Child Care and Education	1
Korean Journal of Thinking Development	1
Teacher Education Research	1
International Journal of Early Childhood Education	1
Korea Practice Association for Early Childhood Education	1
The Korean Journal of Community Living Science	1
CNU Journal of Educational Studies	1
Journal of The Korean Society of Mathematical Education Series A: The Mathematical Education	1
Journal of the Korean Society for Computer Game	1
The Association of Young Children Studies	1

CHAPTER 11

EDUCATING FOR THE FUTURE: AN OUTSIDER'S VIEW OF SOUTH KOREA MATHEMATICS EDUCATION

Lillie R. Albert

Lynch School of Education, Boston College
140 Commonwealth Avenue, Chestnut Hill, Massachusetts 02467
E-mail: albertli@bc.edu

This chapter provides observations about challenges and opportunities regarding the strengths of South Korea mathematics education program. The challenges and opportunities discussed are aligned with findings of an ongoing study of Korean elementary teachers' mathematical knowledge for teaching mathematics. The major argument is that current reform-based instruction is to educate for the future and not the present. Three qualities of the current reform movement support this argument: retooling teaching and learning to develop creativity, problem solving and communication.

1 Introduction

When engaging in discussions concerning mathematics teaching and learning regarding Korean education, a general conclusion is South Korea education system is one of the best in the world. However, this conclusion is too often made based on limited information gathered from South Korea's rankings on international assessments, such as PISA. This conclusion is framed within misconceptions about teaching and learning in South Korean classrooms. For example, some believe that because learning competition is intense, Korean students spend their time exclusively studying and never having the opportunity to relax or engage in extra curricular activities. Another partisan commentary centers on the belief that from grades first to twelfth, 45 to 50 students are crammed into classrooms, requiring little student-to-student interactions in which the predominate teaching method is a lecture and the primary learning approach is memorization. Too often, these perspectives are groundless, lacking any empirical or observational evidence.

This chapter outlines observations of challenges and opportunities regarding the strengths of South Korea mathematics education program. These challenges and opportunities discussed are aligned with findings of an ongoing study of Korean elementary teachers' mathematical knowledge for teaching mathematics. In this chapter, I put forth the argument that current reform-based instruction is to educate for the future and not the present. I support this argument by focusing on three qualities of the current reform movement: retooling teaching and learning to develop creativity, problem solving and communication. To understand the challenges of educating for the future in the first section of this paper, I describe how my learning experiences as a researcher and educator shape my views of mathematics education in South Korea as an outsider. Next, I discuss the qualities of educating for the future through the lens of curriculum and instruction, learning communities, and professional development.

1.1 *The Outsider Context*

A little over twenty years ago while studying for my doctoral degree in mathematics education, I met Jae-sook, a Korean student who was also studying for her doctoral degree; she introduced me to Korean culture, customs and traditions. Because I had spent my final years as a middle school mathematics teachers instructing mostly students that had immigrated with their parents to the United States from Vietnam, Laos and Cambodia, my attentiveness to South Asian culture, particularly education, had become an integral part of my desire to study and understand aspects of teaching and learning for children in these and other Southeastern Asian countries. Thus, after meeting Jae-sook, I quickly developed a relationship with her and continued that relationship for many years to follow. In fact, we started our careers as university professors at the same university. Also, through the years, I continued to read books and engage in conversations with others who had lived in or visited Korea. The continuing development of the Internet provided another avenue for exploring Korean history and culture, especially the

South Korean Education System. It was during this time that I met several mathematics professors at the National Council of Teachers of Mathematics Annual Conference. I was offered an opportunity to review a set of mathematics textbooks and to write the forward for them, which lead to an invitation to visit South Korea. Finally, in late spring of 2010, I visited South Korea for two weeks and during that time; I gave a series of lectures, visited an elementary classroom and observed an afterschool program for young children. Since then, I have visited South Korea, twice more and have future visits scheduled. These visits have provided me with opportunities to spend time in elementary school classrooms, observing mathematics instruction, interviewing teachers and analyzing lesson plans developed by these teachers. In addition, during my last visit in 2013, I spent time at a South Korean high school. These experiences have shaped my view of mathematics education in Korea.

Although I have been in South Korean elementary schools off and on during the past several years, I do not believe myself to be an *insider*, since I have not spent sufficient uninterrupted days in these schools to fully understand all of their complexities. Neither do I think of myself a true *outsider*, since I had spent adequate time interacting with a diverse number of Korean educators, scholars and students, including a few policymakers and curriculum developers. These substantiated interactions and encounters have provided me opportunities to become familiar with South Korean Education System's organization, structure, and educational goals. While during the majority of my visits I did not interact directly with students, I observed many different types of classrooms from grades one to six, and I have had the opportunity to observe mathematics and technology lessons in a high school. I have seen classes in which the students were engaged in the learning process, working in small groups as well as classes in which students did not appeared to be engaged and interactions were limited between teacher and students. I believe the fundamental elements of these experiences informing my view of Korean education are varied, yet, rich in context, culture, and innovation; furthermore, the depth of these experiences serve as a framework for the following sections of this chapter.

2 Curriculum and Instruction

Before beginning a mathematics lesson, a teacher will take time to consider the mathematics curriculum. Schiro contends that a curriculum philosophy "embodies beliefs about the type of knowledge that should be taught in schools, the inherent nature of children, what school learning consists of, how teachers should instruct children, and how children should be assessed."[32] Thus, a curriculum encompasses far more than content and the textbook alone, but rather, the theories behind what is taught, the action of implementing instruction, and the experiences of students as a consequence[21, 32]. These qualities may emphasize applicable curriculum standards, what topics are to be taught and their sequence, which texts and technologies are to be used, which real-world connections can be highlighted, and how the current concepts or skills being taught extend from those of previous instruction.

2.1 *Pedagogical Content Knowledge for Teaching Mathematics*

What also must be considered among all of these qualities is the teacher's ability to implement mathematical practices that effectively integrate subject matter knowledge with pedagogical knowledge, which is referred to as *pedagogical content knowledge*[33, 34] and later described by Ball as *pedagogical content knowledge for teaching mathematics*[6]. "Pedagogical content knowledge addresses how to teach mathematics content and how to understand students' thinking. This includes taking into consideration both the cultural background of the students as well as their preferences for various teaching and learning styles"[3]. This conception of pedagogical content knowledge assumes that there is a balance between pedagogy and content, which according to Shulman includes some features of pedagogical knowledge, e.g., knowledge of curriculum, knowledge of learners or knowledge about assessment and evaluation processes.

A major assertion that also is relevant to Shulman's conception is that curriculum knowledge is supported by both content knowledge and knowledge of assessment processes, while pedagogical knowledge is supported by both knowledge of learners and knowledge of assessment processes[26]. This conception is well documented in mathematics teaching in Korean elementary classrooms. When planning mathematics lesson, elementary teachers report considering theoretical perspectives such as Bruner's discovery learning theory as well as considering their students' mathematical background. This quality included not only how knowledgeable students are regarding concepts and skills but also students' mathematics attitude. For example, one teacher stated, "First, students' mathematical attitude is more important than their mathematics knowledge. Second, the attitude is not a personal attitude, but it is the desire to discover. Third, students desire to make mathematics discoveries contributes to a positive classroom atmosphere." This teacher added that essential to the classroom atmosphere are teaching and learning practices whereby students perform mathematics experiments that include mathematics playing. Thus this teacher considers specific mathematics knowledge of the learner as well as the learner's content disposition and how these specific elements of knowledge contributed to knowledge of the classroom-learning context that is unique or specific to the learner.

2.2 Creativity and Problem Solving

Creative thinking and problem-solving skills are used not only in mathematics but also in all subject areas. As such, developing these skills is essential to life-long learning. The NCTM Standards highlight the many benefits that effective problem-solving skills have on student learning:

By learning problem solving in mathematics, students should acquire ways of thinking, habits of persistence and curiosity, and confidence in unfamiliar situations that will serve them well outside the mathematics classroom. In everyday life and in the workplace, being a good problem solver can lead to great advantages. Problem solving means engaging in a task for which the solution method is not known in advance. In order to find a solution, students must draw on their knowledge and through this process; they will often develop new mathematical understanding[27].

The elementary school classroom is an excellent place in which to engage students in mathematical learning experiences necessary to achieve many of these goals. Because mathematics concepts build on each other, constant contemplation and practice using them on a regular basis is essential to internalizing these concepts. Using problem solving activities to develop creativity will help students keep mathematics on the forefront of their minds, making it easier for them to recall important concepts when transitioning from a pure skill/memorization approach to an approach that values and focuses on process[1].

For students to engage successfully in problem solving tasks, the teacher should provide initial guidance. Put differently, when students are at the beginning stage of learning how to solve problems, the teacher takes the lead as a model and guide. For example, in a Korean fifth grade classroom, the students were studying the properties of a rectangular prism and the task introduced by their teacher required them to compare the properties between a rectangular prism and a cube. During their comparison activity, they also needed to discuss the relationship between the two three-D shapes. The teacher approached this activity by first showing a picture of a rectangular shaped cake, offering students information about the shape but not revealing the properties; thus, the first step was to help students understand the information presented in the problem.

The teacher led a discussion that encourages students to explore and think through the underlying mathematical structure of the problem instead of focusing on extraneous, surface details of the problem. The focus of the discussion was on process, modeled by the teacher, which illustrates systematic thinking. Thus, the teacher showed students how to approach, plan, and think about the problem. Then students placed on their desk various sizes boxes that they had brought from their home. First, students working alone examined their boxes; second, working in pairs they compared their boxes; third, working in small groups they engage in debate about the properties of the boxes; and then, they concluded the lesson by engaging in whole class discussion. The teacher continued the modeling by demonstrating to students how to develop explanations that described what they know about the rectangular prisms and cubes and what they need to find out about them. For instance, one student stated that the rectangular prism was "a box-shaped object." The teacher asked the student to explain further and the student said that it was "sort of like the cube." Then, another student stated that both "the cube and the rectangular prism have six sides and all right angles." This part of the instruction and discussion highlighted the process and made the steps explicit that students might think about or go through in constructing meaning of the problem in their own words. Once the teacher was assured that students understand the problem, students were encouraged to continue to explore the boxes, discussing and debating the various properties that became evidence through their exploration.

The teacher continues instruction by asking probing questions and by giving helpful hints that scaffold student learning. Once students have solved the problem, the teacher engages them in a discussion in which students review the solution strategy they applied to the problem and evaluate the reasonableness of the solution obtained to see if it makes sense in meeting the goal of the problem solving task. Through exploration with real boxes, discussion of the differences and similarities between cubes and rectangular prisms, and debating conjectures made by each others, students in this class were able to conclude that rectangular prisms are recognizable objects found in everyday life, such as household appliances, buildings, and boxes on grocery store shelves. In addition, they were able to see that the two three-D shapes have several similarities but only one difference, and all cubes are rectangular prisms but not all rectangular prisms are cubes. With the support of their teacher, the students were able reason that the one distinction between cubes and rectangular prisms is the shape of the six faces. Each face of a cube is a square, and all of the squares are of equal size. Each face of a rectangular is a rectangle, and a minimum of four of the rectangles is the same.

The importance of the reasoning behind process that students take on during problem solving is often underestimated. It is this process, not the end product, that is most important to understanding mathematics on an abstract, conceptual level. Through problem solving and discovery, students will be challenged to ask and explore *why* various mathematical properties are true. When students reach their own conclusions through their own methods of solving problems, they come to a deeper understanding of the content learned[10]. In addition, the best way to develop the deepest, most thorough understanding of any concept is to discuss and debate idea with others. Thus, sharing the problem solving and creative processes students developed to reach their conclusions will solidify and enhance their learning experience.

In South Korea educating for the future embraces problem solving and creativity. This is certainly an appropriate way to transform school mathematics that has been associated with performing computations and learning isolated facts with little time devoted to reasoning and communication skills. The development of creative and critical thinking and problem solving skills requires learners to engage in activities in which they must identify and solve various problems and then communicate the thinking process they used to arrive at possible solutions. Educating for the future means that teachers engage students in an analysis of problem solving methodology, which they can then apply to content-focused as well as real-world problems. In particular, in the classrooms in which I conducted observations, students were actively engaged in the process for solving problems and explaining their thinking. The analysis of observation and interview data indicated that the South Korean mathematics education program in the elementary grades is aimed in the right direction.

3   The Learning Community: The Role of Communication and Technology

South Korea reform-based instruction calls for mathematics classroom learning communities that include communication practices that are deliberate and intentional. In the study of mathematics instruction in Korean classrooms, I found that the communication practices I observed mirror the perspectives of Bakhtin, Bruner and Vygotsky regarding the role of language, i.e., verbal, oral, and written, in learning and development[5,11,38,39]. The process of learning is both individual and sociocultural, the use of language is crucial in helping both the individual and the collective make sense of experiences. The teachers I observed used language that scaffolded or mediated their *actions* as they participated in mathematical practices with their students. Action is emphasized here because Vygotsky and Luria strongly suggested that action is continuously linked to communication. If we exclude the inflated feature of language ("the word"), then what surfaces is an "underestimation of volitional [preference or desire] action, action in its highest forms, that is, action tied to the word,"[40] which plays a major role in the development of higher cognitive functions (i.e., evaluating the reasonableness of a solution).

A pertinent argument is that the teacher's actions and communication within classroom learning contexts may be a better source for understanding the relationship that exists between mathematics teaching and learning. Bakhtin's notion of the influence of speech on word meaning has some bearing on this argument[5]. Bakhtin proposed that the channel of communication "must not be separated from the realm of discourse, that is, from language as a concrete integral phenomenon. Language is only in the dialogic [or conversation] interaction of those who make use of it"[5]. Therefore, actual dialogic interaction must be grounded in the relationship that permeates social discourse with people and must not be separated from it. As Vygotsky proposed, "all the higher functions originate as actual relations between [people]"[39]. The major argument rests on the assumption that development cannot be separated from social contexts or from language, verbal, oral or written. Therefore, the starting point for the students was geared toward having them engage in talk, task, and content that they were familiar with – the traditional nature of mathematics teaching and learning. However, the intention was to move their understanding and thinking forward to a higher intellectual level about the content. Language, Vygotsky and Luria maintained, along the way becomes "intellectualized and develops on the basis of *action*, lift this action to a supreme level...If *at the beginning* of development there stands the act, independent of the word, then at the end of it there stands the word which becomes the act, the work which makes [individual's] action free"[40] (p. 170, emphasis in original).

3.1 *Language Communication in the Learning Community*

In South Korean mathematical learning communities, when the teacher applied deliberate pedagogical practices for learning and understanding mathematics content, understanding took on personal and intellectual qualities for the students. The teacher, as the knowledgeable other scaffolded learning and understanding, gradually allowing the students to monitor and regulate their own learning of the material, decide the appropriateness of different strategies, and successfully complete given problems, independently or collectively[2]. For example, during the interview, one teacher explained his communication actions in this way, "First, I let students discuss or debate with each other, and then I provide a mathematical explanation. Second, the students are the owner of the classroom. So, thirty percent of the communication time is teacher-to-students, forty percent is student-to-student and thirty percent is a mixtures of teacher-to-student and student-to-student." When asked why debates and discussions are important, the teacher responded, "They are key to helping students find mathematical misconceptions and at the same time helping them develop their knowledge and understanding." Other teachers that I interviewed shared similar statements about the importance of mathematics communication. Another finding is that in effective learning communities verbal interactions that occur between students when they engage in collaborative activities are just as important as interactions between a teacher and a student.

Because of its very social nature, collaborative learning allows for an exchange of dialogue between students that is unlike anything possible in the whole-class or individual settings. Gillies notes that because the exchanges in this setting promote verbal interaction among students, they therefore fit into the Vygotskian scheme of "verbal interactions as a catalyst for promoting thinking"[17]. Students engage in productive verbal interactions when they give help to their peers, justify their answers or their strategies in response to their peers' challenges, and work with one another to resolve disagreements and discrepancies[17]. Then, for South Korea, a direction for professional development as we aim toward educating for the future is preparing teachers how to think critically about the verbal interactions that occur in group learning situations or between them and their students. In doing so, teachers should begin to consider ways through which to promote effective verbal interaction and to discourage ineffectual verbal interaction. When discussing verbal interaction, it is important to note that communication is not merely limited to speaking; rather watching and listening play an equally important part of the communication process in collaborative problem solving[1,2]. Watching and listening are not always passive behaviors; however they must occur in the context of balanced behavior[4].

When considering verbal interactions in terms of its contribution to student learning, researchers first deem whether a particular comment initiates discussion of a new topic or if it responds to another students' initiation by continuing an ongoing discussion. A comment is an initiation if it focuses the groups' attention to discuss a new or related topic or if it poses a question on a new aspect of the problem. Questions are further categorized as being specific questions (either towards specific individuals or about a specific topic) or general questions. It follows naturally that asking specific task-related questions correlate to student learning more strongly than does asking general questions. Other forms of initiation may include statements of confusion, reports of errors in thinking or calculation.

Furthermore, detailed explanations posed by students in responding to critical questions emphatically relates with learning in the collaborative environments. The extent to which a student elaborates her ideas by thoroughly explaining them to her peers correlates positively with increased student achievement. Students learn little from short terminal responses, such as providing a one-word answer to another's question; however, the explanations students receive from each other are critical to learning in groups[16,17,41]. Perhaps these critical elaborations have such an effect on other students' learning because they model an efficient thought process, when given correctly. In explaining one's thinking to a peer, that student scaffolds the problem solving ability of another student. According to Gillies, these explanations must be "timely, relevant to the recipient's need for help, correct, and of sufficient detail to enable them to correct any misunderstanding"[16]. With respect to constructing mathematical understanding, there is a monumental difference between receiving the answer to a question and receiving an explanation as to why that particular answer is correct. Thus, Korean teachers need to engage in professional development activities that provide experiences about how to work with their students to help them realize that to succeed in the collaborative setting they must not only ask clear and specific questions, but that they must also give elaborate responses that trigger mathematical understanding and recall[41].

3.2 *Knowledge Transformation through Learning Technologies*

New technology-based resources such as iPads or tablets are on the path to becoming a significant component of mathematics education programs. Learning to use technology is important not only because technology provides a tool for problem solving, but also because technology has become a natural part of our daily lives. Well-developed e-learning applications are seen as promoting active participation of students, assisting them in the development of critical and creative thinking and encouraging students to contemplate their ideas as well as those presented by their classmates[12,24]. For example, the introduction of iPads and tablets into the classroom is timely in that they offer practical applications in electronic formats that can be augmented, modified or redefined to enhance and transform mathematics teaching and learning[22,23,25].

A useful way to think about knowledge transformation through learning technologies is to contemplate how to use them to leverage pedagogical content knowledge for teaching mathematics. When you create something with technology, it changes human capacity in which the goal is to not just deliver content but to also use it to make sense and develop meaning of it. Therefore, the inclusion of learning technologies with mathematical knowledge and practices may provide students opportunities to apply the mathematical knowledge and values they acquire in their classroom to other contexts. A four-level framework developed by Puentedura and known as the SAMR Model (Substitution-Augmentation-Modification-Redefinition) provides an illustration of effective technology integration in the classroom[29,30,31] (See Figure 11-1.). Substitution and augmentation are considered enhancement levels, illustrating standard or conventional use of technology while the transformation levels of the modification and redefinition focus on the reinvention or alteration of how technology used in the classroom. According to Puentedura, the transformation levels are not superior to the enhancement levels; thus the overarching goal is to use learning technologies in ways that best promote students learning[31].

At the *substitution* level, technology acts as a direct tool with no functional change. A practical example is a PDF scan of mathematics textbook or word-processor software used similar to a typewriter. However, at the *augmentation* level, the technology acts as a tool with functional improvement. In this case, the PDF copy of a mathematics textbook might come with hyperlinks that connect the user to web related sources. The next level of use is *modification*, which allows for considerable redesign of tasks. At this level, the e-book would include integrated activities with functions that allow the user to manipulate as well as interact with mathematics applications. The *redefinition* level is the highest level in which the technology allows for the construction of new tasks, such as a mathematics e-book that allows the user to construct or extend mathematics apps, write instruction for their use, and then share the information online with others. The intention is that an e-book should not be a PDF scan of the textbook. Digital textbooks need to provide a fundamentally different experience for the learner. Digital technologies, such as e-books need to include interactive objects, figures,

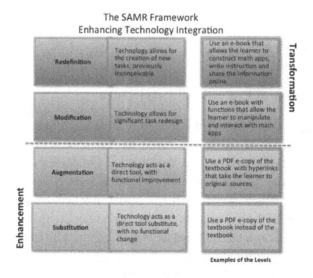

Figure 11-1. The SAMR Framework (adapted from Puentedura31)

photos, diagrams and videos, contains a fast, fluid navigation system, and allows highlighting, note taking, and lesson reviews. All of these aspects allow teachers to increase the instructional value of a lesson and to offer content that may help their students personalize the learning process. According to the research, a critical element that assists students in personalizing learning relies on how the teacher models use of learning technologies, such as tablets[12].

In South Korean classrooms, teachers used technology to model various mathematical concepts and to place concepts into real world contexts via a short video clip or an animation; students' use of learning technologies was limited. As South Korea move to become the first country in the world to digitalize all of its mathematics textbooks, the next step for those who are working on this development is to consider how the new e-textbooks will go beyond the substitution level but contain elements that engage teachers as well as their students in ways that modify or redefine mathematical learning. In summary, educating for the future must deliver opportunities for students to study mathematics through problem solving explorations, skills development, and technology resources to provide reinforcement and enrichment of challenging mathematical content. Then, students will learn to make connections between abstract ideas and their use in a range of academic and real-world situations.

4   The Role of Professional Development

Research has confirmed that classrooms exemplifying problem solving and collaborative grouping cultivate positive effects on students' mathematical disposition and learning[8,9,37]. Fundamental to this finding is the notion that problem solving and collaborative work need to engage teachers and their students in a rigorous intellectual process in which making sense of mathematics content is pertinent to their lives. In a study to measure content knowledge for teaching, Hill and Ball findings suggest, "Teaching mathematics requires an appreciation of mathematical reasoning, understanding the meaning of mathematical ideas and procedures, and knowing how ideas and procedures connect"[19]. They conclude, citing supporting research, that an essential way to influence the teaching of mathematics in classrooms is through quality professional development activities that focus on mathematical knowledge for teaching[13,18]. Furthermore, the strength of this conclusion is that in order to develop the pedagogical skills necessary to convey mathematics in this way, the teachers need professional development experiences that will provide them exposure to learning in this manner. Teaching mathematics through problem solving and reasoning requires educators to challenge how they think about mathematics and break with some procedural routines in order to advance student learning and creativity.

4.1 *The Professional Learning Community*

Darling-Hammond and her colleagues summarize their research in a report on the status of professional development in the U.S. and abroad[14]. Their in-depth analysis was used to establish standards for measuring growth in professional development over a period of time. The following outcomes are essentially relevant to the insights presented in this section: "Sustained and intensive professional development for teachers is related to student achievement gains. Effective professional development is intensive, ongoing, and connected to practice and focuses on the teaching and learning of specific academic content"[14]. When examining data relative to experiences of teachers, the authors found that most teachers do not have access to professional development that meet these criteria and when teachers engage in professional learning activities, they are most likely to report that content specific activities are useful and find other areas to be of less value to them as teachers[14].

The South Korea government supports teachers' professional development and the lack of appropriate access does not seem to be the circumstance for most South Korean teachers. Although access is available, participation in professional development is considered to engage about forty percent of it teachers in such activities[20]. However, in some cases, Korean teachers develop courses for professional development activities. For example, the eleven teachers that participated in my study on pedagogical content knowledge for teaching mathematics at the elementary developed an assessment course. The major purpose of this course was for the teachers to work together to construct assessment items that can be used in their classrooms, to be involved in professional development opportunities over a period of time, and to engage in deep discussion and reflection about mathematics teaching and student learning.

What are the next steps needed that will enhance teachers' pedagogical content knowledge for teaching mathematics, advance students' mathematical understanding, and integrate learning technologies? Emerging from the literature is the claim that the hallmark of a professional development is the focus on teacher and student learning, collaboration, and accountability[7,14,15,24,35,36]. Then, a goal of those responsible in South Korea for establishing and maintaining professional development learning communities must do this with the support of school leaders and teachers themselves. Such communities must employ that teachers learn to work collaboratively to advance pedagogical practices, improve student learning and performance, and hold themselves responsible for learning outcomes. These features, argues DuFour, require members of the learning community to organize their learning around three essential elements: what mathematics students need to learn, what assessment indicators suggest that students have learned, and how to address the needs of students who are struggling to learn mathematics[15].

5   Closing Thoughts

First, to effectively implement innovative components of instruction, such as knowledge of new learning technologies, what must also be integrated among all of these considerations discussed in this chapter are teachers and their students' disposition toward the subject. This includes, but is by not limited to, the enjoyment of mathematics and the process of learning and doing mathematics that encapsulates the belief that mathematics is a human activity used in everyday life[28].

Second, innovative components should also promote student development of deeper understanding of mathematics concepts as well as engage students in investigating advanced mathematics concepts using a process of discovery through real-world problems that encapsulate students' mathematical knowledge, while promoting critical and creative thinking. Thus, if one of the goals of educating for the future is student learning, then this latter task may be the most critical for an educator and may outweigh many, if not all, of the more procedural duties of the educator's job. As Korean policymakers, curriculum developers and teachers embark upon educating for the future, they need to work as a collective, developing an understanding of the importance of sharing and researching ideas, activities, and materials. Therefore, all stakeholders must engage in processes that improve mathematical teaching and learning for all students.

References

1. Albert, L. R. (2000). Outside In, Inside Out: Seventh grade students' mathematical thought processes. *Educational Studies in Mathematics*, 41, 109-142.
2. Albert, L. R. with Corea, D. &, Macadino, V. (2012). *Rhetorical ways of thinking: Vygotskian theory and mathematical problem solving.* New York: Springer Publishing Company.
3. An, S., Kulm, G., &, Wu, Z. (2004). The pedagogical content knowledge of middle school, mathematics teachers in China and the U.S. *Journal of Mathematics Teacher Education,* 7, 145-172.
4. Artzt, Alice and Armour-Thomas, Eleanor (1992). Development of a Cognitive-Metacognitive Framework for Protocol Analysis of Mathematical Problem Solving in Small Groups. *Cognition and Instruction*, 9(2), 137-175.
5. Bakhtin, M. (1984). *Problems of Dostoevsky's poetics.* Minneapolis: University of Minnesota Press.
6. Ball. D. L. (1993). With an eye on the mathematical horizon: Dilemmas of teaching elementary school mathematics. *Elementary School Journal,* 93(4), 373-397.
7. Banilower, E. R., Boyd, S. E., Pasley, J. D., Weiss, I. R. (2006) *Lessons from a decade of mathematics and science reform: A capstone report for the local systemic change through teacher enhancement initiative.* Chapel Hill, NC: Horizons Research.
8. Boaler, J. (2008). Promoting relational equity and high mathematics achievement through an innovative mixed-ability approach. *British Educational Research Journal*, 34(2), 167–194.
9. Boaler, J. (2010). Stories of success: Changing students' lives through sense making and reasoning. In Strutchens, M. E., & Quander, J. R. (Eds.), *Focus in high school mathematics: Fostering reasoning and sense making for all students* (pp. 1-27). Reston, VA: National Council of Teacher of Mathematics.

10. Bruner, J. (1967). *On Knowing: Essays for the Left Hand.* Boston: Harvard University Press.
11. Bruner, J. (1987) Prologue. *In L. Vygotsky, The collected works of L. S. Vygotsky.* (Cole, M., Scribner, S., John-Steiner, V. & Souberman, E. Trans.) Cambridge, MA: Harvard University Press.
12. Cochrane, T. (2010 October). *An m-learning journey: Mobile web 2.0 critical success factors.* Paper presented at the M-Learning 2010: The 9th International Conference on Mobile Learning, Valletta, Malta.
13. Cohen, D. K., & Hill, H. (2001). *Learning policy: When state education reform works.* New Haven, CT: Yale University Press.
14. Darling-Hammond, L., Wei, R. C., Andree, A., Richardson, N., & Orphanos, S., (2009). *Professional learning in the learning professional: A status report on teacher development in the United States and abroad.* National Staff Development Council, Standard University.
15. DuFour, R. (2004). What is a "professional learning community"? *Educational Leadership, 61*(8), 6-11.
16. Gillies, Robyn M. (2004). The effects of cooperative learning on junior high school students during small group learning. *Learning and Instruction,* 14, 197-213.
17. Gillies, Robyn M. (2006*).* Teachers' and students' verbal behaviors during cooperative and small-group learning. *British Journal of Educational Psychology, 76, 271-287.*
18. Hill, H. C. (2004). Professional development standards and practices in elementary school mathematics. *Elementary School Journal,* 104, 345-363.
19. Hill, H. C., & Ball, D., (2004). Learning mathematics for teaching results from California's mathematics professional development institutes. *Journal for Research in Mathematics Education,* 35, 330-351.
20. Kim, J., Colen, Y. S., & Colen, J. (2012). Reform-based instruction in Korea: Looking over its promises to discover its successes. In J. Kim, I. Han, M. Park, & J. Lee (Eds.) *Mathematics education in Korea: Curricular and teaching and learning practices* (pp.104-127). Singapore: World Scientific Publishing Company.
21. Kliebard, H. (2004). *The struggle for the American curriculum 1893-1958,* 3rd edition. New York: NY: Routledge.
22. Laurillard, D. (2007). Pedagogical forms for mobile learning. In N. Pachler (Ed.), *Mobile learning: Towards a research agenda.* London: WLE Centre, Institute of Education.
23. Manuguerra, M. & Petocz, P. (2011). Promoting student engagement by integrating new technology into tertiary education: The role of the iPad. *Asian Social Science* 7(11), 61-65.
24. Martin, S., & Vallance, M. (2008). The impact of synchronous inter-networked teacher training in information and communication technology integration. *Computers and Education,* 51, 34-53.
25. Melhuish, K. & Falloon, G. (2010). Looking to the future: M-learning with the iPad.

Computers in New Zealand Schools, 22(3), 1-16.

26. Morine-Dershimer, G. & Kent, T. (1999). The complex nature and sources of teachers' pedagogical knowledge. In J. Gess-Newsome and N.G. Lederman (Eds.) *PCK and science education* (pp. 21-50). Netherlands, Kluwer Academic Publishers.

27. National Council of Teachers of Mathematics. (2000). *Principles and standards for school mathematics*. Reston, VA: National Council of Teachers of Mathematics.

28. National Research Council. (2001). *Adding it up: Helping children learn mathematics*. J. Kilpatrick, J. Swafford, and B. Findell (Eds.). Mathematics Learning Study Committee, Center for Education, Division of Behavioral and Social Sciences and Education. Washington, DC: National Academy Press.

29. Peng, H., Su, Y., Chou, C. & Tsai, C. (2009). Ubiquitous knowledge construction: mobile learning re-defined and a conceptual framework. *Innovations in Education and Teaching International*, 46(2), 171–183.

30. Puentedura, R. (2006). *Transformation, technology and education*. Retrieved from http://hippaus.com/resources/tte/purntedura_tte.pdf.

31. Puentedura, R. (2008, December 22) *TPCK and SAMR: Models for enhancing technology integration*. Retrieved rom https://itunes.apple. Com/us/itune-u/as-we-may-teach-educational/id380294705.

32. Schiro, M. S. (2008). *Curriculum theory: Conflicting visions and enduring concerns*. Thousand Oaks, CA: Sage Publications.

33. Shulman, L. S. (1986). Paradigms and research programs in the study of teaching: A contemporary perspective. In M. C. Wittrock (Ed.), *Handbook of research on teaching* (3rd edition, pp. 3-36). New York: Macmillan Publishing Company.

34. Shulman, L. (1987). Knowledge and teaching: Foundations of the new reform. *Harvard Educational Review*, 57(1), 1-22.

35. Sowder, J. T. (2007). The mathematical education and development of teachers. In F. K. Lester Jr. (Ed.), *Second handbook of research on mathematics teaching and learning* (pp. 157-223). Charlotte, NC: Information Age.

36. Sparks, D. (2005). Leading for transformation in teaching, learning, and relationships. In R. DuFour, R. Eaker, & R. DuFour (Eds.), *On Common Ground: The power of professional learning communities* (pp. 155-175). Bloomington, IN: National Education Service.

37. Steffero, M. (2010). *Tracing belief and behaviors of participant in a longitudinal study for the development of mathematical ideas and reasoning*. Doctoral Dissertation Rutgers, the State University of New Jersey.

38. Vygotsky, L. S. (1978). *Mind in society: The development of higher psychological processes*. Cambridge, MA: Harvard University Press.

39. Vygotsky, L. S. (1994). The problem of the environment. In R. Van Der Veer and J. Valsiner (Eds.), *The Vygotsky reader* (pp. 338-354). Cambridge, Massachusetts: Blackwell.

40. Vygotsky, L. S. & Luria, A. (1994). Tool and symbol in child development. In R. Van Der Veer and J. Valsiner (Eds.), *The Vygotsky reader* (pp. 99-174). Cambridge, Massachusetts: Blackwell.

41. Webb, Noreen M., Farviar, Sydney (1994). Promoting helping behavior in cooperative small groups in middle school mathematics. *American Educational Research Journal*, 31, 396-395.

CHAPTER 12

THE COMMON CORE MATHEMATICS STANDARDS AND IMPLICATIONS FOR THE SOUTH KOREAN CURRICULUM

Yong S. Colen

Department of Mathematics, Indiana University of Pennsylvania
307 Stright Hall, 210 South 10th Street, Indiana, PA 15705
E-mail: yscolen@iup.edu

Jung Y. Colen

Department of Curriculum and Instruction (Mathematics Education)
The Pennsylvania State University
113 Chambers Building, University Park, PA 16802
E-mail: jyc125@psu.edu

Jinho Kim

Department of Mathematics Education
Daegu National University of Education
#219 Jungang-daero, Namgu, Daegu City, 705-715, South Korea
E-mail: jk478kim@dnue.ac.kr

This chapter examines the meticulous process in crafting the Common Core Mathematics Standards and the final product, the Standards for Mathematical Content and Practice. The focus then shifts to the impetuses behind this undertaking. The final section explores some plausible implications for the South Korean Mathematics Curriculum. These range from allocating resources for meaningful professional development for teachers and translating the South Korean school mathematics textbooks to eliciting input from the greater community.

1 Curriculum Envisioned by the Common Core Mathematics Standards

This section examines the ideals of the Common Core Mathematics Standards. In particular, both the process and the product convey a comprehensive, systematic approach to improving the U.S. mathematics standards.

1.1 *The Vision of the Common Core State Standards Initiative*

In December 2008, the National Governors Association (NGA), the Council of Chief State School Officers (CCSSO), and Achieve, Inc. published *Benchmarking for Success: Ensuring U.S. Students Receive a World-Class Education*[1]. In this report, governors, state education chiefs, and prominent education researchers articulated five overarching action items. Principally, Action 1 recommended: "Upgrade state standards by adopting a common core of internationally benchmarked standards in math and language arts for grades K-12 to ensure that students are equipped with the necessary knowledge and skills to be globally competitive."[2]

In the following year, governors and state education chiefs, representing 48 states, two territories, and the District of Columbia and through their membership in the NGA Center for Best Practices and CCSSO, launched the Common Core State Standards Initiative (CCSSI)[3]. The subtitle of this initiative, "Preparing America's Students for College and Career," clearly designated the dual focus. Explicitly, "state school chiefs and governors recognized the value of consistent, real-world learning goals and launched this effort to ensure all students, regardless of where they live, are graduating high school prepared for college, career, and life."[4]

1.2 *The Mathematics Work Team*

In June 2009, the Mathematics Work Team[5] was charged to develop the K-12 standards. The team members represented a highly qualified cross-section within the mathematics education community. Out of 51 members, a sample is listed below:

William McCallum, Lead
Head, Department of Mathematics
The University of Arizona
Senior Consultant, Achieve, Inc.

Deborah Loewenberg Ball
Dean, School of Education
University of Michigan

Sybilla Beckmann
Professor, Department of Mathematics
University of Georgia

Douglas H. Clements
SUNY Distinguished Professor
Department of Learning and Instruction
The State University of New York at Buffalo

Francis (Skip) Fennell
Professor, Department of Education
McDaniel College
Past-President, NCTM

Sol Garfunkel
Executive Director
COMAP (The Consortium for Mathematics and Its Applications)

Roger Howe
William Kenan Jr. Professor, Department of Mathematics
Yale University

Becky Pittard
National Board Certified Teacher
Pine Trail Elementary School, Florida

Andrew Schwartz
Assessment Manager, Research and Development
The College Board

Donna Watts
Coordinator, Mathematics and STEM Initiatives
Maryland State Department of Education

Hung-Hsi Wu
Professor Emeritus, Department of Mathematics
University of California-Berkeley

1.3 *The Mathematics Feedback Group*

The Mathematics Feedback Group[6] was constituted to review the K-12 standards. The members' professional achievements are both notable and diverse. The below represents a sample from 22 members:

Richard Askey
Professor Emeritus, Department of Mathematics
University of Wisconsin-Madison

Hyman Bass
Samuel Eilenberg Distinguished University Professor
Department of Mathematics and School of Education
University of Michigan

Andrew Chen
President, EduTron Corporation

Scott Eddins
Tennessee Mathematics Coordinator
President, Association of State Supervisors of Mathematics

Roxy Peck
Associate Dean and Professor, College of Science and Mathematics
California Polytechnic State University, San Luis Obispo

Ronald Schwarz
High School Math Instructional Specialist
Department of Science, Technology, Engineering and Mathematics
Office of Curriculum, Standards and Academic Engagement
New York City Department of Education

Uri Treisman
Professor, Departments of Mathematics and Public Affairs
Executive Director, Charles A. Dana Center
The University of Texas at Austin

W. Stephen Wilson
Professor, Department of Mathematics
Johns Hopkins University

1.4 *The Standards-Development Principles*

The CCSSI development team used the below standards-development principles[7] to guide their deliberations and formulations of the standards:

Goal: The standards, as a whole, must be essential, rigorous, clear and specific, coherent, and internationally benchmarked.

Essential: The standards must be reasonable in scope in defining the knowledge and skills students should have to be ready to succeed in entry-level, credit-bearing, academic college courses and in workforce training programs.

Rigorous: The standards will include high-level, cognitive demands by asking students to demonstrate deep, conceptual understanding through the application of content knowledge and skills to new situations. High-level, cognitive demand includes reasoning, problem solving, synthesis, analysis, and justification.

Clear and specific: The standards should provide sufficient guidance and clarity so that they are teachable, learnable, and measurable. The standards will also be clear and understandable to the general public. Quality standards are precise and provide sufficient detail to convey the level of performance expected without being overly prescriptive (the "what" and not the "how"). The standards should maintain a relatively consistent level of grain size.

Teachable and learnable: Provide sufficient guidance for the design of curricula and instructional materials. The standards must be reasonable in scope, instructionally manageable, and promote depth of understanding. The standards will not prescribe *how* they are taught and learned but will allow teachers flexibility to teach and students to learn in various, instructionally relevant contexts.

Measureable: Student attainment of the standards should be observable and verifiable, and the standards can be used to develop broader assessment frameworks.

Coherent: The standards should convey a <u>unified vision</u> of the big ideas and supporting concepts within a discipline and reflect a <u>progression</u> of learning that is meaningful and appropriate.

Grade-by-grade standards: The standards will have limited repetition across the grades or grade spans to help educators align instruction to the standards.

Internationally benchmarked: The standards will be informed by the content, rigor, and organization of standards of high-performing countries so that all students are prepared for succeeding in our global economy and society.

1.5 *Input from the Greater Community*

In June 2010, NGA and CCSSO released the final version of the Common Core State Standards (CCSS). In addition to the feedback group's comments, the work team received invaluable input from the participating states' departments of education, the national education organizations, the public, and the Validation Committee.

First, since the states were the ones who would eventually adopt and implement the standards, the work team elicited their comments at various stages. Second, the National Education Association (NEA), the American Federation of Teachers (AFT), and the National Council of Teachers of Mathematics (NCTM) were integrally involved in bringing together teachers who provided specific and constructive feedback on the standards. Third, almost 10,000 individuals participated in the online survey, and among the respondents, 48% identified themselves as "K-12 teachers," 20% as "parents," 6% as "administrators," 5% as "post secondary faculty members or researchers," and 2% as "students."[8] Some of the more important findings and feedback were:

- A majority of respondents valued having common education standards across the states.

- At least three-fourths of educators, from pre-kindergarten through higher education, reacted positively or very positively to each of the general topics.

- Without explicitly saying so, educators gravitated toward a curriculum that revisits concepts and topics across grades.

- Many found the document's language difficult to follow.

- Respondents also indicated an interest in seeing examples for each standard in order to ensure understanding of the expectations.

- Many respondents' concerns were focused on the implementation of the standards. The majority of respondents were comfortable with the quality of the standards, but they wanted to be sure that enough would be done to ensure successful implementation. They desired for the standards to exist as part of a well-supported, cohesive, seamless, education system.

- Parents were concerned that "children are being pushed too hard to meet standards at too early an age.... It is too much to ask 5, 6 and even some 7 year olds to sit at a desk and learn all these things. We need to let our young children be young children."

Finally, the 25-member Validation Committee[9], composed of the leading figures in the education standards community, provided an independent validation of the process in identifying the CCSS. Specifically, the committee:

- reviewed the process by which evidence was used to create K-12 and college- and career-readiness standards;

- determined that the standards-development principles were adhered to by examining the standards for:
  - a grounding in available evidence and research,
  - evidence of the knowledge and skills that students need in order to be college- and career-ready,
  - a proper level of clarity and specificity,
  - evidence that the standards are comparable with other leading countries' expectations.

1.6 *The Common Core State Standards for Mathematics*

The work team describes the Common Core State Standards for Mathematics (CCSSM)[10] as "what students should understand and be able to do in their study of mathematics" and "clear signposts along the way to the goal of college and career readiness for all students."[11] Optimistic in achieving this goal, forty-four states have adopted the CCSSM as of June 2014. Curriculum envisioned by the CCSSM comprises the following distinct features.

1.6.1 *The Standards for Mathematical Content*

The Standards for Mathematical Content, in particular the mathematical progressions (or sequenced topics), is focused, coherent, mathematically sound, and based on research evidence. In all, there are 11 standards by domain:

1. Counting and Cardinality
2. Operations and Algebraic Thinking
3. Number and Operations in Base Ten
4. Number and Operations-Fractions
5. Measurement and Data
6. Geometry
7. Ratios and Proportional Relationships
8. The Number System
9. Expressions and Equations
10. Functions
11. Statistics and Probability

For example, the "Number and Operations-Fractions" domain[12] spans grades three to five. The overarching cluster theme at grade three is: **Develop understanding of fractions as numbers.** The first standard states:

> Understand a fraction $1/b$ as the quantity formed by 1 part when a whole is partitioned into b equal parts; understand a fraction a/b as the quantity formed by a parts of size $1/b$.

At grade four, the overarching cluster themes and their respective, sample standards progress to:

Extend understanding of fraction equivalence and ordering.
Explain why a fraction a/b is equivalent to a fraction $(n \times a)/(n \times b)$ by using visual fraction models with attention to how the number and size of the parts differ even though the two fractions themselves are

the same size. Use this principle to recognize and generate equivalent fractions.

Build fractions from unit fractions.
Understand a fraction a/b with $a > 1$ as a sum of fractions $1/b$.

Understand decimal notation for fractions and compare decimal fractions.
Express a fraction with denominator 10 as an equivalent fraction with denominator 100 and use this technique to add two fractions with respective denominators 10 and 100. *For example, express 3/10 as 30/100 and add 3/10 + 4/100 = 34/100.*

At grade five, the overarching cluster themes and their respective, sample standards progress to:

Use equivalent fractions as a strategy to add and subtract fractions.
Add and subtract fractions with unlike denominators (including mixed numbers) by replacing given fractions with equivalent fractions in such a way as to produce an equivalent sum or difference of fractions with like denominators. For example, 2/3 + 5/4 = 8/12 + 15/12 = 23/12. (In general, a/b + c/d = (ad + bc)/bd.)

Apply and extend previous understandings of multiplication and division.
Interpret a fraction as division of the numerator by the denominator (a/b = a ÷ b). Solve word problems involving division of whole numbers leading to answers in the form of fractions or mixed numbers, e.g., by using visual fraction models or equations to represent the problem. For example, interpret 3/4 as the result of dividing 3 by 4, noting that 3/4 multiplied by 4 equals 3, and that when 3 wholes are shared equally among 4 people each person has a share of size 3/4. If 9 people want to share a 50-pound sack of rice equally by weight, how many pounds of rice should each person get? Between what two whole numbers does your answer lie?

The above sample standards and the mathematical progression of the identified domain across grades are a testament that the CCSSM provides the clarity and specificity that were long overdue and are welcoming.

1.6.2 *The Standards for Mathematical Practice*

The CCSSM delineates the Standards for Mathematical Practice that mathematics teachers should strive to develop in all students. These mathematical practices are the refinements from

- the NCTM's Process Standards[13]: problem solving, reasoning and proof, communication, connections, and representation;

- the National Research Council's *Adding It Up* and its Strands of Mathematical Proficiency[14]: conceptual understanding, procedural fluency, strategic competence, adaptive reasoning, and productive disposition.

The eight Standards for Mathematical Practice[15] and some of their attributes are:

1. **Make sense of problems and persevere in solving them.**
   Mathematically proficient students start by explaining to themselves the meaning of a problem and looking for entry points to its solution. They analyze givens, constraints, relationships, and goals. They make conjectures about the form and meaning of the solution and plan a solution pathway rather than simply jumping into a solution attempt. They consider analogous problems and try special cases and simpler forms of the original problem in order to gain insight into its solution. They monitor and evaluate their progress and change course if necessary... Mathematically proficient students check their answers to problems using a different method, and they continually ask themselves, "Does this make sense?" They can understand the approaches of others to solving complex problems and identify correspondences between different approaches.

2. Reason abstractly and quantitatively.

Mathematically proficient students make sense of quantities and their relationships in problem situations. They bring two complementary abilities to bear on problems involving quantitative relationships: the ability to *decontextualize*—to abstract a given situation and represent it symbolically and manipulate the representing symbols as if they have a life of their own without necessarily attending to their referents—and the ability to *contextualize*—to pause as needed during the manipulation process in order to probe into the referents for the symbols involved. Quantitative reasoning entails habits of creating a coherent representation of the problem at hand; considering the units involved; attending to the meaning of quantities, not just how to compute them; and knowing and flexibly using different properties of operations and objects.

3. Construct viable arguments and critique the reasoning of others.

Mathematically proficient students understand and use stated assumptions, definitions, and previously established results in constructing arguments. They make conjectures and build a logical progression of statements to explore the truth of their conjectures. They are able to analyze situations by breaking them into cases and can recognize and use counterexamples. They justify their conclusions, communicate them to others, and respond to the arguments of others. They reason inductively about data and make plausible arguments that take into account the context from which the data arose. Mathematically proficient students are also able to compare the effectiveness of two plausible arguments, distinguish correct logic or reasoning from that which is flawed, and, if there is a flaw in an argument, explain what it is. Elementary students can construct arguments using concrete referents such as objects, drawings, diagrams, and actions. Such arguments can make sense and be correct even though they are not generalized or made formal until later grades. Students at all grades can listen or read the arguments of others, decide whether they make sense, and ask useful questions to clarify or improve the arguments.

4. Model with mathematics.

Mathematically proficient students can apply the mathematics they know to solve problems arising in everyday life, society, and the workplace. In early grades, this might be as simple as writing an addition equation to describe a situation. In middle grades, a student might apply proportional reasoning to plan a school event or analyze a problem in the community. By high school, a student might use geometry to solve a design problem or use a function to describe how one quantity of interest depends on another. Mathematically proficient students, who can apply what they know, are comfortable making assumptions and approximations to simplify a complicated situation and realize that these may need revision later. They are able to identify important quantities in a practical situation and map their relationships using such tools as diagrams, two-column tables, graphs, flowcharts, and formulas... They routinely interpret their mathematical results in the context of the situation, reflect on whether the results make sense, possibly improve the model if it has not served its purpose.

5. Use appropriate tools strategically.

Mathematically proficient students consider the available tools when solving a mathematical problem. These tools might include pencil and paper, concrete models, a ruler, a protractor, a calculator, a spreadsheet, a computer algebra system, a statistical package, or dynamic geometry software. Proficient students, who are sufficiently familiar with tools appropriate for their grade or course, make sound decisions about when each of these tools might be helpful and recognize both the insight to be gained and their limitations. When making mathematical models, they know that technology can enable them to visualize the results of varying assumptions, explore consequences, and compare predictions with data.

6. Attend to precision.

Mathematically proficient students try to communicate precisely to others. They try to use clear definitions in discussion with others and in their own reasoning. They state the meaning of the symbols they choose, including using the equal sign consistently and appropriately.

They are careful about specifying units of measure and labeling axes to clarify the correspondence with quantities in a problem. They calculate accurately and efficiently and express numerical answers with a degree of precision appropriate for the problem context. In the elementary grades, students give carefully formulated explanations to each other. By the time they reach high school, they have learned to examine claims and make explicit use of definitions.

7. **Look for and make use of structure.**

Mathematically proficient students look closely to discern a pattern or structure. Young students, for example, might notice that three and seven more is the same amount as seven and three more, or they may sort a collection of shapes according to how many sides the shapes have. Later, students will see 7×8 equals $7 \times 5 + 7 \times 3$ in preparation for learning about the distributive property. In the expression $x^2 + 9x + 14$, older students can see the 14 as 2×7 and the 9 as $2 + 7$. They recognize the significance of an existing line in a geometric figure and can use the strategy of drawing an auxiliary line for solving problems. They can see complicated things, such as some algebraic expressions, as single objects or as being composed of several objects. For example, they can see $5 - 3(x - y)^2$ as 5 minus a positive number times a square and use that to realize that the original value cannot be more than 5 for any real numbers x and y. They can also step back for an overview and shift perspective.

8. **Look for and express regularity in repeated reasoning.**

Mathematically proficient students notice if calculations are repeated and look both for general methods and for shortcuts. Upper elementary students might notice when dividing 25 by 11 that they are repeating the same calculations over and over again and conclude they have a repeating decimal. By paying attention to the calculation of slope as they repeatedly check whether points are on the line through $(1, 2)$ with slope 3, middle school students might abstract the equation $(y - 2)/(x - 1) = 3$. Noticing the regularity in the way terms cancel when expanding $(x - 1)(x + 1)$, $(x - 1)(x^2 + x + 1)$, and $(x - 1)(x^3 + x^2 + x + 1)$ might lead students to the general formula for the sum of a geometric series. As mathematically proficient students work to solve a problem, they maintain oversight of the process, attend to the details, and continually evaluate the reasonableness of their intermediate results.

It is fitting to end the section with the following thoughts:

Expectations that begin with the word "understand" are often especially good opportunities to connect the practices to the content. Students who lack understanding of a topic may rely on procedures too heavily.... The points of intersection [between the Standards for Mathematical Content and the Standards for Mathematical Practice] are intended to be weighted toward central and generative concepts in the school mathematics curriculum that most merit the time, resources, innovative energies, and focus necessary to qualitatively improve the curriculum, instruction, assessment, professional development, and student achievement in mathematics.[16]

These Standards are not intended to be new names for old ways of doing business. They are a call to take the next step. It is time for states to work together to build on lessons learned from two decades of standards based reforms. It is time to recognize that standards are not just promises to our children, but promises we intend to keep.[17]

2   Impetuses behind the Creation of the CCSSM

The CCSSM is a testament to the inadequacy of the U.S. school mathematics curriculum. This section highlights some impetuses behind the creation of the CCSSM.

2.1 *Poor Performances on the International and National Assessments*

It is no coincidence that the work team articulated the standards-development principle:

Internationally benchmarked: The standards will be informed by the content, rigor, and organization of standards of high-performing countries so that all students are prepared for succeeding in our global economy and society.

So which countries were "high-performing"? To answer this question, we shall examine two international mathematics assessments. The data have provided valuable insights on the U.S. children's knowledge of mathematics, the dissimilar opportunities to learn mathematics in the classroom, and the disparity of textbook quality.

Since 1995, the Third International Mathematics and Science Study and the subsequent, Trends in International Mathematics and Science Study (both referred to as TIMSS) have provided reliable and timely data on the mathematics and science achievement of the participating countries.[18] The U.S. Department of Education, National Center for Education Statistics (NCES) has served as the administrator.

In 2011 (the most recent year with available data), more than 20,000 U.S. students, representing more than 1,000 school districts and joining almost 500,000 other students around the world, partook in the assessment study. Among the fourth graders from the 57 participating countries and territories[19], the average mathematics score of U.S. students was 541, a result higher than the TIMSS scale average of 500. Overall, they placed 15th. Eight countries had statistically higher averages, and the top five scoring countries were: Singapore (with 606 average), South Korea (605), Hong Kong (602), Chinese Taipei (591), and Japan (585). Within the U.S., North Carolina obtained the highest score of 554. Over the years, there has been a positive trend line of the TIMSS average mathematics scores for U.S. fourth graders: In 1995, the score was 518; in 2007, 529; and in 2011, 541.

At grade eight, 56 countries participated. The average mathematics score of U.S. students was 509. Overall, they placed 24th, and 11 countries had statistically higher averages. Almost mirroring the above list, the top five countries with the average mathematics scores above the U.S. were: South Korea (613), Singapore (611), Chinese Taipei (609), Hong Kong (586) and Japan (570). Within the U.S., Massachusetts garnered the highest score of 561. Finally, the TIMSS average mathematics scores for U.S. eighth graders in 1995 was 492; in 2007, 508; and in 2011, 509.

At the secondary school level, the 1995 TIMSS findings depicted an alarming picture. Among the 21 participating countries, 14 statistically outperformed U.S. twelfth graders on the Mathematics General Knowledge portion. The average U.S. score was meager 461, and the two countries that scored significantly lower were Cyprus (446) and South Africa (356). Compared to 45% of international students, only 32% of U.S. students were able to answer the below item (see Figure 12-1) correctly.[20]

Stu wants to wrap some ribbon around a box as shown below and have 25 centimeters left to tie a bow.

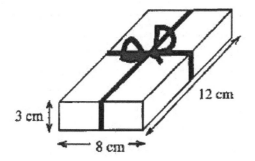

How long a piece of ribbon does he need?

A. 46 cm
B. 52 cm
C. 65 cm
D. 71 cm
E. 77 cm

Figure 12-1. Mathematics General Knowledge Item

What is more unsettling is the fact that in the Advanced Mathematics Performance portion, U.S. students scored 442, and no other country scored significantly lower. Furthermore, according to the TIMSS findings—the score of 541 for U.S. fourth graders, 509 for U.S. eighth graders, 461 and 442 for U.S. twelfth graders—and compared to their international cohorts, gains in U.S. students' knowledge of mathematics appear to diminish as they progress through the diverse school systems.

The planned 2015 TIMSS at the specified grades (fourth, eighth, and twelfth) will furnish an opportunity to compare to the existing baseline.

Another international assessment that measures students' mathematics literacy is the Program for International Student Assessment (PISA).[21] Since 2000 and every three years, the Organization for Economic Cooperation and Development (OECD) has coordinated PISA, and like TIMSS, the NCES has administered the assessment. As 15-year-old students near the end of compulsory schooling, PISA measures their mathematics, reading, and science literacy and emphasizes cross-curricular competencies, such as problem solving and "functional skills." In the most recent PISA in 2012, compared to the OECD average mathematics score of 494, the U.S. scored 481. The top five performing countries were: Shanghai (613), Singapore (573), Hong Kong (561), Chinese Taipei (560), and South Korea (554).

Since 1969, the National Assessment of Educational Progress (NAEP)[22] has measured U.S. students' knowledge of mathematics, reading, science, writing, the arts, civics, economics, geography, and U.S. history. Collected every two years from fourth, eighth, and twelfth graders, NAEP data have provided useful information for state-to-state comparisons and year-to-year progress. While there is danger in highlighting a selective finding, Wearne and Kouba underscored that only 38 percent of U.S. eighth graders, given five choices, had correctly selected the solution to a 15 percent tip on a typical meal.[23]

American Mathematical Association of Two-Year Colleges' *Crossroads in Mathematics: Standards for Introductory College Mathematics before Calculus* and *Beyond Crossroads: Implementing Mathematics Standards in the First Two Years of College* have documented U.S. high school graduates' lack of mathematical foundation for post-secondary study.[24, 25] There are real consequences to relatively poor preparation of students in mathematics and science as attested by the passage of the law:

On September 24, 1998, the U.S. House of Representatives approved a bill allowing an additional 142,500 foreign skilled workers to enter the country, thus exempting them from normal immigration quotas. The bill was designed to meet the needs of high-technology industries that are unable to find adequate skilled workers among the U.S. population.[26]

2.2 Lack of Focused, Coherent, and Mathematically Sound Standards

At the "Summit on Mathematics" in 2003, Schmidt addressed the audience in part:

> What does a focused and rigorous curriculum look like in the top achieving countries [on TIMSS]? The number of topics that children are expected to learn at a given grade level is relatively small, permitting a thorough and deep coverage of each topic. For example, nine topics are the average number intended in the second grade. The U.S. by contrast expects second grade teachers to cover twice as many mathematics topics. The result is a characterization of the U.S. curriculum as *a mile wide and an inch deep* (emphasis added).[27]

The first recommendation from *Foundations for Success: The Final Report of the National Mathematics Advisory Panel* echoed the above analysis:

> A focused, coherent progression of mathematics learning, with an emphasis on proficiency with key topics, should become the norm in elementary and middle school mathematics curricula. Any approach that continually revisits topics year after year without closure is to be avoided.[28]

In particular, the panel recommended that "The Benchmarks for the Critical Foundations" should consist of "whole numbers," "fractions," and particular aspects of "geometry and measurement."[29] Likewise, compared to the K-6 mathematics standards in the U.S., Ginsburg, Leinwand, and Decker found the composite standards of Singapore, South Korea, and Hong Kong focused on the number, measurement, and geometry strands and much less on data analysis and algebra.[30]

As part of the 1995 TIMSS, researchers examined the videos of eighth-grade, mathematics lessons from Germany, Japan, and the U.S.

Based on the subsequent report, *The TIMSS Videotape Classroom Study*[31], Stigler and Hiebert concluded:

> An initial indication of the learning opportunities for students in a mathematics lesson is the nature and level of mathematics that is on the playing field. What is in the lesson, substance-wise, for the students to use to construct mathematical knowledge? We have learned that, in this regard, U.S. students are at a disadvantage. They encounter mathematics that is at a lower level, is somewhat more superficial, and is not as fully or coherently developed as the mathematics encountered by their German and Japanese peers... Indeed, when the coded data are examined, differences in the content of the lessons appear to be even larger than when individual lessons are compared. U.S. students encounter less-challenging mathematics, and because it is presented in a less-coherent way, they must work harder to make sense of it than their peers in Germany and Japan.[32]

In grade eight, the following CCSSM standard exemplifies what it means to be "focused," "coherent," and "mathematically sound":

> Use similar triangles to explain why the slope m is the same between any two distinct points on a non-vertical line in the coordinate plane.[33]

Developing this concept well, mathematics teachers will provide a sound justification as to why *any two distinct points* on a non-vertical line will determine its slope.

2.3 *Diverse Curricula due to States Sovereignty and Local Autonomy*

Transition from the CCSSM to a CCSSM-aligned curriculum will be challenging. In the U.S., states sovereignty and district-level autonomy to formulate school curricula are very much valued and protected. For example, Pennsylvania Department of Education (PDE) has compiled *Academic Standards for Mathematics: Grades Pre K-High School*[34]. Omitting any reference to the CCSSM, the adopted standards are called, *The PA Core Standards*. One can clearly infer that this was a

Pennsylvania initiative. More importantly, the degree of wasted resources in regurgitating the CCSSM and producing a PDE document that is anything-but-useful reflects the state's pride and to some degree, a lack of professional judgment to recognize the needs of the schools, teachers, and students. In short, utility of such document is difficult to discern.

Another factor that compounds the challenge in developing a focused, coherent, and mathematically sound curriculum is the PDE's directive: "Although the standards are not a curriculum or a prescribed series of activities, school entities will use them to develop a local school curriculum that will meet local students' needs."[35] Due to 14,178 school districts[36] throughout the U.S., formulating and implementing CCSSM-based curricula will be a daunting task. To date, the states that adopted the CCSSM have not made a concerted, collaborative effort to preserve the "Common" standards.

Academic freedom entrenched within the U.S. culture appears to be a hindrance toward academic success. For example, during the 1950s and 1960s, there were about three dozen, reformed-oriented projects in addition to the School Mathematics Study Group (SMSG). Davis recollects:

These projects aimed and moved in different (and sometimes irreconcilable) directions. Some retained the traditional grade-level placement of topics, whereas others altered it drastically. Some focused on pure mathematics, others on uses of mathematics; and some combined mathematics so completely within their overall content structure that one could not identify it as a separate ingredient. Some focused on an abstract introduction of mathematics, whereas others sought to make child's early experiences with mathematics more concrete. Some retained traditional forms of classroom activities, whereas other projects sought to introduce learning experiences of entirely different sorts. Some sought to influence the curriculum mainly by writing textbooks; others chose instead to work directly with teachers, by offering workshops of in-service courses, or even by creating "teacher centers" where teachers could come and learn the new goals and methods. Some projects even moved into classrooms, working alongside teachers as they tried out new approaches.[37]

Incredulously, he observes: "Most schools seem hardly to have been affected at all [by these programs]... It was possible at the time to walk into almost any school in the United States and see mathematics teaching that was little different from typical teaching before World War II."[38]

2.4 *Textbook School Mathematics*

Quality of the school mathematics textbooks is particularly important in the U.S. Freeman and Porter found that most mathematics teachers rely heavily on the textbook.[39] In 1973, Ed Begle, director of the SMSG (the program that became synonymous with the "new math" and the "reform"), wrote: "The curriculum ... clearly does influence what student learn.... In fact, it seems at present to be the only variable that on the one hand we can manipulate and on the other hand does affect student learning."[40]

In both traditional and non-traditional, U.S. school mathematics textbooks, Ginsburg and colleagues determined weak development and mechanical presentation of mathematical concepts.[41] Steen provides a further caution regarding curricula that "distort mathematics."[42] Wu claims that since the demise of the "new math" in the early 1970s, U.S. schools have had "a *de facto* national mathematics curriculum, namely, the curriculum dictated by school mathematics textbooks."[43] He has coined this curriculum, "Textbook School Mathematics" (TSM)[44]. Specifically, Wu asserts:

> Until we eradicate TSM from the school curriculum, any mathematical standard that calls for the teaching of a mathematical topic in a certain grade will do nothing but rearrange the mathematically flawed presentations in TSM.[45]

> The resulting [TSM] curriculum distorts mathematic in the sense that it often withholds precise definitions and logical reasoning, fails to point out interconnections between major topics such as whole

numbers and fractions, and employs ambiguous language that ultimately leads to widespread non-learning.[46]

For example, Wu advocates that students should initially examine fractions as "part-whole" numbers as outlined in the previously-mentioned, CCSSM cluster, "Develop understanding of fractions as numbers." In his effort, he provides the following definition of fractions:

Let k, l be whole numbers with $l > 0$. Divide each of the line segments [0, 1], [1, 2], [2, 3], [3, 4], … into l segments of equal length. These division points together with the whole numbers now form an infinite sequence of equally spaced markers on the number line (in the sense that the lengths of the segments between consecutive markers are equal to each other). The first marker to the right of 0 is by definition $1/l$. The second marker to the right of 0 is by definition $2/l$, the third $3/l$, etc., and the k-th is k/l. The collection of these k/l's for all whole numbers k and l, with $l > 0$, is called the *fractions*. The number k is called the *numerator* of the fraction k/l, and the number l its *denominator*.[47]

There are two typical reactions from the mathematics education community: (1) This coherent and mathematically-sound progression extends students' prior understanding of the number system and offers a contrast to most elementary school mathematics textbooks that provide students three distinct meanings of fractions: part-whole, quotient, and ratio. (2) This "mathematically-correct and yet detached from the classroom reality" approach does not provide opportunities for conceptual learning. In his defense, upon a careful examination of Wu's "Part 2. Fractions" in *Understanding Numbers in Elementary School Mathematics*[48], one gets a strong sense that he is committed in *developing* both teachers' and students' understanding of mathematics.

Wu, a professor emeritus of mathematics, embodies the past and contemporary mathematicians, like Max Beberman, Ed Begle, Henry Pollak, Hyman Bass, Sybilla Beckmann and William McCallum, who have devoted much energy to improve the quality of U.S. mathematics education. In reform efforts, their voices are needed and should be welcomed.

Investigating the quality of U.S. textbooks, McClintick scathingly concluded in "The Great American Textbook Scandal":

> In the intensely lobbied textbook selection process in states like California, intellectual content takes a back seat to salesmanship, political correctness, self-esteem for the students and the need to dumb-down lessons so that one product can capture a large market. In the U.S., textbooks sell well if they are designed to hold the attention of children accustomed to MTV and spiffy Internet graphics; they don't sell well if they are challenging.[49]

Reflecting on the "CCSSM-aligned" mathematics textbooks, Keeghan informs the readers in "Afraid of Your Child's Math Textbook? You Should Be":

> There may be a reason you can't figure out some of those math problems in your son or daughter's math text and it might have nothing at all to do with you. That math homework you're trying to help your child muddle through might include problems with no possible solution. It could be that key information or steps are missing, that the problem involves a concept your child hasn't yet been introduced to, or that the math problem is structurally unsound for a host of other reasons.[50]

In 2013–2014, while guiding mathematical learning of a fifth-grader during the after-school, tutorial sessions, we have had opportunities to examine her textbook, *Prentice Hall Mathematics Common Core, Course 1*[51]. The textbook's numerous errors of all types strongly substantiate the findings by Keeghan.

Annually, U.S. elementary and secondary textbook publishing is a four-billon dollar industry.[52] Based on the input from textbook selection committees, school boards often use the following criterion in adopting mathematics textbooks: Among the available textbook series, which one is least worse.[53] Moreover, mathematics teachers face a dilemma: To ascertain the quality of a textbook, they need to use it for a year, but this requires a district-wide adoption—a commitment to use the textbook, irrespective of its quality, for the next four, five, or six years.

Coxford recalls that by the late 1960s and early 1970s, the reform ideas represented by SMSG were in retreat, and most commercial textbooks had implemented "an interpretation of the SMSG curriculum."[54] At present, major textbook publishing companies have rushed to publish, sell, and make profit on "CCSSM-aligned" mathematics textbooks, and this may come with a great cost: Like the demise of the SMSG efforts due to poor quality textbooks, the CCSSM may face the similar fate.

2.5 *Woefully Inadequate Teacher Preparation Programs and School Culture*

It is an understatement to say that U.S. mathematics teachers, as a whole, face many obstacles to teach well. The first obstacle is the teachers' inadequate knowledge of school mathematics. Ball, Lubienski, and Mewborn concluded that "studies of teachers' mathematical knowledge––elementary and secondary, preservice and experienced—reveal pervasive weaknesses in U.S. teachers' understanding of fundamental mathematical ideas and relationships."[55] Ma found that "the knowledge gap between the U.S. and Chinese teachers parallels the learning gap between U.S. and Chinese students.... The real mathematical thinking going on in a classroom, in fact, depends heavily on the teacher's understanding of mathematics."[56] Wilson shares the sentiment: "The field of mathematics education is first of all about mathematics."[57] A study authored by the National Center for Research on Teacher Education concluded that most U.S. teacher preparation programs concentrate on the teaching of mathematics rather than on the mathematics itself.[58] For a successful implementation of a CCSSM-aligned curriculum, Wu advocates for raising the demand of teachers' content knowledge.[59] Hill, Rowan, and Ball specifically recommend "efforts to improve teachers' mathematical knowledge through content-focused professional development."[60]

In the Conference Board of the Mathematical Sciences volume, *The Mathematical Education of Teachers*, the first recommendation states:

> *Prospective teachers need mathematics courses that develop a deep understanding of the mathematics they will teach.* The mathematical knowledge needed by teachers at all levels is substantial, yet quite different from that required by students pursuing other mathematics-related professions. Prospective teachers need to understand the fundamental principles that underlie school mathematics, so that they can teach it to diverse groups of students as a coherent, reasoned activity and communicate an appreciation of the elegance and power of the subject. With such knowledge, they can foster an enthusiasm for mathematics and a deep understanding among their students. College courses developing this knowledge should make connections between the mathematics being studied and mathematics prospective teachers will teach.[61]

The second obstacle lies in the teachers' inadequate pedagogical knowledge. A National Advisory Committee on Mathematical Education's study on school mathematics instruction determined that in elementary schools, "teachers are essentially teaching the same way they were taught in school. Almost none of the concepts, methods, or big ideas or modern mathematics has appeared."[62] Stigler and Hiebert further observed:

> Teaching is a complex system... It has now been documented in several studies that teachers asked to change features of their teaching often modify the features to fit within their pre-existing system instead of changing the system itself. The system assimilates individual changes and swallows them up. Thus, although surface features appear to change, the fundamental nature of the instruction does not."[63]

Before the U.S. House of Representatives' Committee on Economic and Educational Opportunities, in a candid testimony on the issue of inadequate mathematics teacher preparation programs, Albert Shanker declared:

> Doctors don't try to figure out a new technique or procedure for every patient who comes into their office; they begin by using the standard techniques and procedures based on the experience of many doctors over the years. Nobody considers this a way of doctor-proofing medicine, although they do have a name for the failure to use standard practices—it's *malpractice*. The standard practices that all doctors (and other professionals) use contain the wisdom of the profession."[64]

Inability to transform the school culture epitomizes the third obstacle. The National Commission on Teaching and America's Future found that "most schools and teachers cannot achieve the goals set forth in new educational standards, not because they are unwilling, but because they do not know how, and the systems they work in do not support them in doing so."[65]

Another major shortcoming to U.S. mathematics education is due to how states certify teachers. Wilson specifically implicates that at the secondary school level, "certification to teach mathematics, whether through regular programs or alternative programs, does not guarantee the appropriate content preparation for teaching or good teaching practices."[66] In a June 2014 update, the Pennsylvania Department of Education's Division of Professional Education and Teacher Quality announced:

> To maintain high standards for content and pedagogy while providing the school districts with more flexibility for assignment, we are permitting teachers with valid instructional certificates to add the Pre K-4 and the 4-8 certificates by passing the required certification assessment."[67]

In short, a middle school teacher, who had specialized in social science and language and having passed the specified assessment, is now deemed qualified to teach mathematics. An unfortunate outcome to producing inadequate mathematics teachers is that "there is in our society, a widespread lack of confidence in teachers."[68]

2.6 *Despite the Funding Commitment toward Education*

U.S. students' poor mathematical achievements, as attested by the TIMSS, PISA, and NAEP findings, have ensued in the backdrop of a strong economy. Measured in GDP per capita[69] and compared to the mathematically top performing countries, the U.S. economy in 2012 was second only to Singapore (see Table 12-1).

Table 12-1. 2012 GDP Per Capita for U.S. and Mathematically Top Performing Countries[70]

Countries	Singapore	United States	Japan	Hong Kong	South Korea
2012 GDP Per Capita	$52,052	$51,749	$46,731	$36,796	$22,590

Additionally, the U.S. expenditure per primary and secondary student, as measured by the percent of GDP per capita, demonstrates a commitment toward education (see Tables 12-2 and 12-3).

Table 12-2. Expenditure Per Primary Student for U.S. and Mathematically Top Performing Countries71

Countries	Japan	South Korea	United States	Hong Kong	Singa-pore
Expenditure Per Primary Student (Percent of GDP Per Capita)	23.7% (2010)	23.3% (2009)	22.1% (2010)	14.7% (2010)	11.2% (2010)

Table 12-3. Expenditure Per Secondary Student for U.S. and Mathematically Top Performing Countries72

Countries	Japan	United States	South Korea	Hong Kong	Singapore
Expenditure Per Secondary Student (Percent of GDP Per Capita)	24.3% (2010)	24.3% (2010)	23.8% (2009)	17.5% (2010)	17.0% (2010)

In sheer numbers, the total U.S. expenditure for public, elementary and secondary school education was $638 billion for the academic year 2009-2010, and this averages out to about $13,000 per student.[73] Unmistakably, it is difficult to claim that lack of monetary investment in education was a reason for U.S. students' poor mathematical performances.

3 Implications of the CCSSM for the South Korean Curriculum

The U.S. has had a constructive influence on the South Korean mathematics curriculum. Reciprocally, in shaping the CCSSM, the CCSSI cited the South Korean mathematics curriculum. With this backdrop, the chapter concludes with some implications of the CCSSM that South Korean policymakers and educators should consider.

3.1 *South Korea's Influence on the CCSSM*

Due to South Korean students' strong performances on TIMSS and PISA, an important outcome has been a worldwide interest in the country's mathematics curriculum, textbooks, etc. Scholarly manuscripts, such as *Informing Grades 1-6 Mathematics Standards Development: What Can Be Learned from High-Performing Hong Kong, Korea, and Singapore?*[74] and "South Korea: A Success Story in Mathematics Education,"[75] further bolster the claim that South Korea has provided a focused, coherent, and mathematically sound education. This achievement is particularly noteworthy after one considers:

When Korea became independent again in 1945, then, the Japanese departure left a huge, deliberately created gap in trained manpower. Illiteracy was still widespread in Korea—the overall illiteracy rate in 1945 as 78 percent—but even more critical was the shortage of teachers and others with a modicum of secondary education.[76]

During the past seven decades, policymakers and educators have refined the national mathematics curricula. CCSSM's specific citation of the South Korean mathematics curriculum as a consulted resource is a notable recognition.[77]

3.2 *U.S. Influence on the South Korean Curriculum*

The history of U.S. influence on the South Korean mathematics curricula dates back to the 1880s. The end of the Chosun Dynasty (1392-1910) signified a period of influx of U.S. Presbyterian and Methodist missionaries, their establishment of schools (e.g., laying the groundwork for Yonsei University in 1885 and Ewha Womans University in 1886), and their transmission of the "western learning." Both Hansung High School and Hansung Girls High School taught mathematics in 1885.[78]

The post WW II period has afforded an expanded influence. While difficult to quantify, many U.S.-educated, Korean nationals have trained generations of mathematics teachers and contributed toward writing textbooks and developing curricula. For instance, South Korea's 3rd National Mathematics Curriculum that was implemented from 1973 to 1981 was very much influenced by the "new math."[79] Moreover, coinciding this initiative, Sorensen concludes:

Serious and sustained special attention to scientific and technical education came only in 1973 with the establishment of vocational schools associated with the "movement to scientificize the whole people" that was developed in conjunction with the government's heavy and chemical industrialization plan begun in the same year.[80]

Also in 1973, mathematician Morris Kline penned *Why Johnny Can't Add: The Failure of the New Math*[81] that outlined the major criticism of the "new math": Children did not learn the basic mathematics skills. This perception served as the catalyst toward the "back to basics" movement in the U.S. and in turn shaped the South Korea's 4th National Mathematics Curriculum (1982-1988).[82] Finally, NCTM's *Curriculum and Evaluation Standards for School Mathematics*[83] and *Principles and Standards for School Mathematics*[84] have had much influence on the subsequent South Korea's National Mathematics Curricula.[85, 86, 87]

3.3 *Examining the Standards for Mathematical Content*

Despite the states' mathematics standards that are based on the past NCTM Standards or the CCSSM, the U.S. does not have a national curriculum. It is also important to acknowledge that adopting certain standards does not necessarily translate into a correlated curriculum or desired classroom practice. The CCSSM forewarns, "the standards themselves do not dictate curriculum, pedagogy, or delivery of content."[88] In short, while it is much easier to assess the U.S. students' mathematical knowledge, depicting the U.S. school mathematics curriculum is a difficult task.

In comparison, the post Korean War, 1st National Mathematics Curriculum (1955–1963) ushered in the South Korean national mathematics curricula. Current research, consideration for students' needs, and the desire for improvement have guided the periodic revisions. Furthermore, South Korea's National Mathematics Curriculum is intrinsically connected to the development of mathematics textbooks.[89] In fact, the Ministry of Education, the Ministry of Education and Human Resources Development, and the Ministry of Education, Science, and Technology oversee the quality of one series of the elementary mathematics textbooks and several versions of the middle and high school mathematics textbooks.[90]

Composing the next National Mathematics Curriculum that includes the textbooks, policymakers and educators should strongly consider incorporating the CCSSM's mathematical progressions found within the domains (see section 1.6.1.). In particular, these mathematical progressions can be easily implemented into the current integrated structure of the South Korean mathematics textbooks.

3.4 *Examining the Standards for Mathematical Practice*

In "Habits of Mind: An Organizing Principle for Mathematics Curricula," Cuoco, Goldenberg, and Mark implore:

A curriculum organized around habits of mind tries to close the gap between what the users and makers of mathematics *do* and what they *say*. Such a curriculum lets students in on the process of creating, inventing, conjecturing, and experimenting... It helps students look for logical and heuristic connections between new ideas and old ones.... A mathematics course must surely contain the results of mathematical thinking, but by organizing the course around the ways of thinking rather than around the results, one gets yet another benefit. Many of the ways of thinking ... bring power and important perspective to domains other than mathematics.[91]

The eight Standards for Mathematical Practice (see section 1.6.2.) afford mathematically focused ways to develop students' habits of mind. South Korean policymakers and educators should commit to accentuating these standards in the next National Mathematics Curriculum. Additionally, developing assessments to measure students' habits of mind will be challenging. However, emerging resources, such as Partnership for Assessment of Readiness for College and Careers (PARCC)[92], the Smarter Balanced Assessment Consortium (Smarter Balanced)[93], the National Center and State Collaborative Partnership[94], and the Dynamic Learning Maps Alternative Assessment System Consortium[95], will serve as important benchmarks.

3.5 *Translating the South Korean School Mathematics Textbooks*

In recent years, the English versions of the Japanese, Russian, and Singapore school mathematics textbooks and workbooks have gained popularity in the U.S. The University of Chicago School Mathematics Project (UCSMP) has translated several Japanese and Russian school mathematics textbooks, and these are readily available.[96] The U.S. editions of the Singapore school mathematics textbooks, workbooks, and teaching resources are used in some school districts, and there is even a companion, *Elementary Mathematics for Teachers*, to train prospective elementary teachers.[97]

Examining school mathematics textbooks is one of the better ways to determine the scope and sequence of the units and the anticipated, learning outcomes. Grow-Maienza and her colleagues translated into English the South Korean elementary school mathematics textbooks based on the 5th National Mathematics Curriculum (1989-1994), and sample chapters are available online.[98] However, the non-textbook, web-based presentations do not fully capture the lessons' simplicity, conciseness, and coherence. South Korean policymakers and educators should explore translating into English and disseminating the national school mathematics textbooks—especially the elementary school series and the companion workbooks and teacher's manuals.

3.6 *Allocating Resources for Meaningful Professional Development for Teachers*

South Korean policymakers and educators should allocate the necessary resources for teachers to implement successfully the envisioned Standards for Mathematical Practice in their classrooms. It is one thing to guide students to "use equivalent fractions as a strategy to add and subtract fractions," and quite another to "make sense of problems and persevere in solving them."

First, the next National Mathematics Curriculum could specify which Standards for Mathematical Practice to emphasize within the units and lessons. This guidance (Points to Consider for Teachers) will develop not only the students' habits of mind, but the teachers' as well. Valuing these pedagogical tools, teachers can meaningfully transform the classroom practice. Second, small-group lesson studies, involving teachers, teacher educators, and curriculum leaders, could provide the collegial support to plan, critique, and modify lessons. Third, in order for students to "construct viable arguments and critique the reasoning of others," they need time to work collaboratively and opportunities to articulate their thoughts. Fewer students in the classrooms and lessened teaching loads will facilitate toward this attainment. Fourth, journal assignments represent another instructional tool for students to reflect on their reasoning. The above list is by no means exhaustive. To develop students' habits of mind, modeled after the CCSSM's Standards for Mathematical Practice, educators may need to start and continue this dialogue.

3.7 *Eliciting Input from the Greater Community*

In the U.S., misinformation about the CCSSM has led to some dissention among teachers, parents, and educators. Teachers are primarily concerned with the lack of needed resources to implement the CCSSM-aligned curricula and the poor assessment outcomes of their students. At times, the general public at large has erroneously concluded that the CCSSM is a federal mandate. For instance, at the website, *Hoosiers Against Common Core*[99], the opponents of the CCSSM list the following reasons[100]:

- Adopting the CCSS takes control of educational content and standards away from parents, taxpayers, local school districts, and states. The CCSS were produced by a closed group and conditionally approved by many states without public review.

- Public education is a state responsibility. It is not the responsibility of the federal government.

- States have had state standards under No Child Left Behind (NCLB) for several years now. There is no evidence from this experience that this allowed students to move from one district to another with minimal interruption of their instructional program. Even with common standards, there will remain wide variances between classrooms, schools, districts, and states. Common standards within states under NCLB did not result in consistency and collaboration among districts within states. Why should we believe the CCSS would bring this about across district and state lines?

- Adoption of the CCSS will result in greater turmoil and confusion for teachers and students. It will result in a loss of learning time and have a negative effect on test results.

- The CCSS represents a massive unevaluated experiment with our students for which they and their parents have been ill informed and have had no opportunity for input.

It is interesting to note that the above list does not comprise any shortcomings of the CCSSM itself.

In his presentation, "The Common Core: Where Do We Go From Here?" Usiskin shares his perceived shortcomings of the CCSSM[101]:

- Overemphasis on paper-and-pencil computation

- Disregard for technology in doing mathematics

- Inadequate treatment of data and statistics in K-5

- Lack of consideration of those going directly from high school to the workplace

Due to the mounting public pressure, Indiana governor acquiesced, and in March 2014, the state became the first one to withdraw formally from the CCSSI. At that time, the governor remarked:

I believe when we reach the end of this process there are going to be many other states around the country that will take a hard look at the way Indiana has taken a step back, designed our own standards and done it in a way where we drew on educators, we drew on citizens, we drew on parents and developed standards that meet the needs of our people."[102]

In retrospect, the CCSSI could have informed the pubic at the onset and throughout the initiative. Nonetheless, section 1.5 (Input from the Greater Community) documents the efforts at great lengths to gather input from the diverse stakeholders.

A monograph by the Korea Foundation for the Advancement of Science and Creativity, *Study of 2009 Revision of the Korean Mathematics Curriculum*, documents some participation from parents and students.[103] South Korea's efforts to elicit input from the greater community are lauded. For the next National Mathematics Curriculum, policymakers and educators should consider an online forum to elicit feedback from teachers, parents, students, administrators, policymakers, mathematics educators, researchers, etc. The standards that are "clear and understandable to the general public" may result in much greater participation. Valuing their voices throughout the drafting stages will be an important step toward the general public's ownership of the resulting curriculum.

In conclusion, the CCSSM is a momentous achievement. Its Standards for Mathematical Content and Practice will shape the U.S. mathematics standards, curricula, and textbooks for many years to come. South Korea can seize this occasion to examine these standards with a critical discernment and modify and adopt some. Garnering public support may pose some challenge, but all stakeholders need to accept the ideal that completing this journey will lead to an improved teaching and learning of school mathematics.

Endnotes

1. National Governors Association (NGA), Council of Chief State School Officers (CCSSO), & Achieve, Inc. (2008). *Benchmarking for success: Ensuring U.S. students receive a world-class education*. Washington, DC: Authors. Available at http://www.nga.org/files/live/sites/NGA/files/pdf/0812BENCHMARKING.P DF.
2. NGA, CCSSO, & Achieve, Inc. (2008). *Benchmarking for success*. (The quote is found on p. 6.)
3. NGA Center for Best Practices, & CCSSO. (n.d.). *Common core state standards initiative: Preparing America's students for college and career*. Available at http://www.corestandards.org.
4. NGA Center for Best Practices, & CCSSO. (n.d.). *Development process*. Retrieved from http://www.corestandards.org/about-the-standards/development -process.
5. NGA Center for Best Practices, & CCSSO. (n.d.). *Common core state standards initiative: K-12 standards development teams*. (The team members can be retrieved from http://www.nga.org/files/live/sites/NGA/files/pdf/2010COMMONCOREK12TEA M.PDF.)
6. NGA Center for Best Practices, & CCSSO. (n.d.). *Common core state standards initiative*. (The group members can be retrieved from http://www.nga.org/ files/live/sites/NGA/files/pdf/2010COMMONCOREK12TEAM.PDF.)
7. NGA Center for Best Practices, & CCSSO. (n.d.). *Common core state standards initiative: Standards-setting criteria*. (These principles can be retrieved from http://www.corestandards.org/assets/Criteria.pdf.)
8. NGA Center for Best Practices, & CCSSO. (n.d.). *Reactions to the March 2010 draft common core state standards: Highlights and themes from the public feedback*. (The cited statistics, as well as the subsequent findings and feedback, can be retrieved from http://www.corestandards.org/assets/k-12-feedback-summary.pdf.)

9. NGA Center for Best Practices, & CCSSO. (2010). *Reaching higher: The common core state standards validation committee.* Washington, DC: Authors. (The committee members and their related work can be retrieved from http://www.corestandards.org/assets/CommonCoreReport_6.10.pdf.)

10. NGA Center for Best Practices, & CCSSO. (2010). Common core state standards for mathematics (CCSSM). Washington, DC: Authors. Available at http://www.corestandards.org/wpcontent/uploads/Math_Standards.pdf.

11. NGA Center for Best Practices, & CCSSO. (2010). *CCSSM.* (The quote is found on p. 4.)

12. NGA Center for Best Practices, & CCSSO. (n.d.). *Number and operations-fractions.* Retrieved from http://www.corestandards.org/Math/Content/NF. (In the following pages, only sample standards are cited.)

13. National Council of Teachers of Mathematics (NCTM). (2000). *Principles and standards for school mathematics.* Reston, VA: Author. (The Process Standards can be found on pp. 52-71.)

14. National Research Council. (2001). *Adding it up: Helping children learn mathematics.* Washington, DC: The National Academies Press. (The Strands of Mathematical Proficiency can be found on pp. 115-56.)

15. NGA Center for Best Practices, & CCSSO. (n.d.). *Standards for mathematical practice.* (The cited mathematical practices can be retrieved from http://www.corestandards.org/Math/Practice.)

16. NGA Center for Best Practices, & CCSSO. (n.d.). *Standards for mathematical practice.* Retrieved from http://www.corestandards.org/Math/Practice.

17. NGA Center for Best Practices, & CCSSO. (n.d*.). Introduction: How to read the grade level standards.* Retrieved from http://www.corestandards.org/Math/Content/introduction/how-to-read-the-grade-level-standards.

18. National Center for Education Statistics (NCES). (n.d.). *Trends in international mathematics and science study (TIMSS).* (The cited statistics can be retrieved from http://nces.ed.gov/TIMSS/index.asp.)

19. *While it is more apt to categorize Hong Kong and Shanghai as territories and not countries, the authors will utilize "country" exclusively henceforth in describing the participating countries and territories.*

20. NCES. (1998). *Pursuing excellence: A study of U.S. twelfth-grade mathematics and science achievement in international context.* Washington, DC: U.S. Government Printing Office. Retrieved from http://nces.ed.gov/pubs98/98049. pdf. (The test item is found on p. 33.)

21. NCES. (n.d.). *Program for international student assessment (PISA).* (The cited statistics can be retrieved from http://nces.ed.gov/surveys/pisa/index.asp.)

22. NCES. (n.d.). *National assessment of educational progress (NAEP).* (The cited statistics can be retrieved from http://nces.ed.gov/nationsreportcard.)

23. Wearne, D, & Kouba, V. (2000). Rational numbers. In E. Silver, & P. Kennedy (Eds.), *Results from the seventh mathematics assessment of the national assessment of education progress* (pp.163-91). Reston, VA: NCTM.

24. American Mathematical Association of Two-Year Colleges (AMATYC). (1995). *Crossroads in mathematics: Standards for introductory college mathematics before*

calculus. Memphis, TN: Author. Available at http://www.amatyc.org/?page=GuidelineCrossroads.

25. AMATYC. (2006). *Beyond crossroads: Implementing mathematics standards in the first two years of college.* Memphis, TN: Author. Available at http://beyondcross_roads.matyc.org/doc/PDFs/BCAll.pdf.

26. Stigler, J., & Hiebert, J. (1999). *The teaching gap: Best ideas from the world's teachers for improving education in the classroom.* New York: The Free Press. (The quote is found on p. 182.)

27. Schmidt, W. (2003). A presentation at the U.S. department of education secretary's summit on mathematics. Retrieved from http://www2.ed.gov/rschstat/research/progs/mathscience/schmidt.html.

28. U.S. Department of Education. (2008). *Foundations for success: The final report of the national mathematics advisory panel.* Retrieved from http://www2.ed.gov/about/bdscomm/list/mathpanel/report/final-report.pdf. (The quote is found on p. xvi.)

29. U.S. Department of Education. (2008). *Foundations for success.* (The quote is found on p. xvii.)

30. Ginsburg, A., Leinwand, S., & Decker, K. (2009). *Informing grades 1-6 mathematics standards development: What can be learned from high-performing Hong Kong, Korea, and Singapore?* Washington, DC: American Institutes for Research. Available at http://www.air.org/sites/default/files/downloads/report/MathStandards_0.pdf.

31. NCES. (1999). *The TIMSS videotape classroom study: Methods and findings from an exploratory research project on eighth-grade mathematics instruction in Germany, Japan, and the United States.* Washington, DC: U.S. Government Printing Office. Available at http://nces.ed.gov/pubs99/1999074.pdf.

32. Stigler, J., & Hiebert, J. (1999). *The teaching gap.* (The quote is found on p. 66.)

33. NGA Center for Best Practices, & CCSSO. (2010). *CCSSM.* (The quote is found on p. 54.)

34. Pennsylvania Department of Education (PDE). (2014). *Academic standards for mathematics: Grades pre k-high school.* Available at http://static.pdesas.org/content/documents/PA%20Core%20Standards%20Mathematics%20PreK-12%20March%202014.pdf.

35. PDE. (2014). *Academic standards for mathematics.* (The quote is found on p. 2.)

36. U.S. Census Bureau. (n.d.). *Public school systems by type of organization and state: 2012.* Retrieved from http://factfinder2.census.gov/faces/tableservices/jsf/pages/productview.xhtml?pid=COG_2012_ORG10.US01&prodType=table.

37. Davis, R. (2003). Changing school mathematics. In G. Stanic, & J. Kilpatrick (Eds.), *A history of school mathematics: Volume 1* (pp. 623-46). Reston, VA: NCTM. (The quote is found on p. 624.)

38. Davis, R. (2003*). Changing school mathematics.* (The quote is found on p. 625.)

39. Freeman, D., & Porter, A. (1989). Do textbooks dictate the content of mathematics instruction in elementary schools? *American Educational Research Journal, 26,* 403-21.

40. Begle, E. (1973). Some lessons learned by SMSG. *Mathematics Teacher, 66,* 207-14. (The quote is found on pp. 207-9.)

41. Ginsburg, A., Cooke, G., Leinwand, S., Noell, J., & Pollock, E. (2005). *Reassessing U.S. international mathematics performance: New findings from the 2003 TIMSS and PISA.* Washington, DC: American Institutes for Research. Available at http://files.eric.ed.gov/fulltext/ED491624.pdf.

42. Steen, L. (2007). Facing facts: Achieving balance in high school mathematics. *Mathematics Teacher, 100,* 86-95. Available at http://www.stolaf.edu/people/steen/Papers/07facing_facts.pdf.

43. Wu, H. (2014). Potential impact of the common core mathematics standards on the American curriculum. In Y. Li, & G. Lappan (Eds.), *Mathematics curriculum in school education* (pp. 119-42). Advances in Mathematics Education, Dordrecht: Springer.

44. Wu, H. (2011). Phoenix rising: Bringing the common core state mathematics standards to life. *American Educator, 35*(3), 3-13. Retrieved from http://www.aft.org/pdfs/americaneducator/fall2011/Wu.pdf.

45. Wu, H. (2014). *Potential impact of the common core mathematics standards.* (The quote is found on p. 6.)

46. Wu, H. (2014). *Potential impact of the common core mathematics standards.* (The quote is found on p. 2.)

47. Wu, H. (2002). *Chapter 2. Fractions (draft).* Retrieved from http://math.berkeley.edu/~wu/EMI2a.pdf. (The quote is found on p. 6.)

48. Wu, H. (2011). *Understanding numbers in elementary school mathematics.* Providence, RI: American Mathematical Society. Available at http://www.amazon.com/dp/0821852604. (See "Part 2. Fractions," pp. 173-374.)

49. McClintick, D. (2000). *The great American textbook scandal.* Retrieved from http://www.forbes.com/forbes/2000/1030/6612178a.html.

50. Keeghan, A. (2012). *Afraid of your child's math textbook? You should be.* Retrieved from http://open.salon.com/blog/annie_keeghan/2012/02/17/afraid_of_your_childs_math_textbook_you_should_be.

51. *Prentice Hall mathematics common core, course 1.* (2013).

52. McClintick, D. (2000). *The great American textbook scandal.*

53. Typically, in the U.S., elected school board members oversee school districts. Examine McClintick's *"The Great American Textbook Scandal"* to get a sense why textbook adoption process in the U.S. is based on the selection of the "least worse."

54. Coxford, A. (2003). *Mathematics curriculum reform: A personal view.* In G. Stanic, & J. Kilpatrick (Eds.), *A history of school mathematics: Volume 1* (pp. 599-621). Reston, VA: NCTM. The quote is found on p. 605.

55. Ball, D., Lubienski, S., & Mewborn, D. (2001). Research on teaching mathematics: The unsolved problem of teachers' mathematical knowledge. *Handbook of research on teaching, 4,* 433-56. (The quote is found on p. 444.)

56. Ma, Liping. (1999). *Knowing and teaching elementary mathematics: Teachers' understanding of fundamental mathematics in China and the United States.* Mahwah, NJ: Lawrence Erlbaum Associates. (The quote is found on pp. 144, 153.)

57. Wilson, J. (2003). A life in mathematics education. In G. Stanic, & J. Kilpatrick (Eds.), *A history of school mathematics: Volume 1* (pp. 1779-807). Reston, VA: NCTM. (The quote is found on p. 1786.)

58. National Center for Research on Teacher Education. (1991). *Findings from the teacher education and learning to teach study: Final report.* East Lansing, MI: Author.

59. Wu, H. (2014). *Potential impact of the common core mathematics standards.* (The quote is found on p. 2.)

60. Hill, H., Rowan, B., & Ball, D. (2005). Effects of teachers' mathematical knowledge for teaching on student achievement. *American Educational Research Journal, 42*(2), 371-406. (The quote is found on p. 400.)

61. Conference Board of the Mathematical Sciences (CBMS). (2001). *The mathematical education of teachers.* Washington, DC: Author. (The quote is found on p. 7.)

62. CBMS. (1975). *Overview and analysis of school mathematics, K-12.* Washington, DC: Author. (The quote is found on p. 77.)

63. Stigler, J., & Hiebert, J. (1999). *The teaching gap.* (The quote is found on p. 98.)

64. Shanker, A. (1997). The October 1995 testimony. *American Educator, 21*(1), 35-6.

65. National Commission on Teaching and America's Future. (1997). *Doing what matters most: Investing in quality teaching.* New York: Author. (The quote is found on p. 1.)

66. Wilson, J. (2003). *A life in mathematics education.* (The quote is found on pp. 1801-2.)

67. The quote is from a June 2014 email communication from the Pennsylvania Department of Education's Division of Professional Education and Teacher Quality.

68. Feiman-Nemser, S., & Floden, R. (1986). The cultures of teaching. In M. Wittrock (Ed.), *Handbook of research on teaching* (pp. 505-26). New York: Macmillan.

69. According to the World Bank, GDP per capita is defined as "gross domestic product divided by midyear population. GDP is the sum of gross value added by all resident producers in the economy plus any product taxes and minus any subsidies not included in the value of the products. It is calculated without making deductions for depreciation of fabricated assets or for depletion and degradation of natural resources." Retrieved from http://data.worldbank.org/indicator/NY.GDP.PCAP.CD.

70. The data are in current U.S. dollars. Comparable data for Chinese Taipei and Shanghai were not available. Retrieved from http://data.worldbank.org/indica tor/NY.GDP.PCAP.CD.

71. The most recent, available data were in 2010, except for South Korea (2009). Comparable datum for Chinese Taipei was not available. Retrieved from http://data.worldbank.org/indicator/SE.XPD.PRIM.PC.ZS/countries.

72. The most recent, available data were in 2010, except for South Korea (2009). Comparable datum for Chinese Taipei was not available. Retrieved from http://data.worldbank.org/indicator/SE.XPD.SECO.PC.ZS/countries.

73. NCES. (n.d.). Retrieved from http://nces.ed.gov/fastfacts/display.asp?id=66.

74. Ginsburg, A., Leinwand, S., & Decker, K. (2009). *Informing grades 1-6 mathematics standards development.*

75. Sami, F. (2013). South Korea: A success story in mathematics education. *MathAMATYC Educator, 4*(2), 22-8. Available at http://c.ymcdn.com/sites/www.amatyc.org/resource/resmgr/educator_feb_2013/sami2013februarymae.pdf.

76. Sorensen, C. (1994). Success and education in South Korea. *Comparative Education Review, 38*(1), 10-35. (The quote is found on pp. 15-6.)
77. NGA Center for Best Practices, & CCSSO. (2010). *CCSSM.* (See p. 91.)
78. Lee, J. (2013). History of mathematics curriculum in Korea. In J. Kim, I. Han, M. Park, & J. Lee (Eds.), *Mathematics education in Korea: Curricular and teaching and learning practices* (pp. 1-20). Hackensack, NJ: World Scientific Publishing.
79. Lee, J. (2013). *History of mathematics curriculum in Korea.*
80. Sorensen, C. (1994). *Success and education in South Korea.* (The quote is found on pp. 10-1.)
81. Kline, M. (1973). *Why Johnny can't add: The failure of the new math.* New York: St. Martin's Press.
82. Lee, J. (2013). *History of mathematics curriculum in Korea.*
83. NCTM. (1989). *Curriculum and evaluation standards for school mathematics.* Reston, VA: Author.
84. NCTM. (2000). *Principles and standards for school mathematics.* Reston, VA: Author.
85. Korea Foundation for the Advancement of Science and Creativity. (2011). *Study of 2009 revision of the Korean mathematics curriculum.* Seoul, South Korea: Author. [In Korean]
86. Hwang, H., & Han, H. (2013). Current national mathematics curriculum. In J. Kim, I. Han, M. Park, & J. Lee (Eds.), *Mathematics education in Korea: Curricular and teaching and learning practices* (pp. 21-61). Hackensack, NJ: World Scientific Publishing.
87. Lee, J. (2013). *History of mathematics curriculum in Korea.*
88. NGA Center for Best Practices, & CCSSO. (2010). *CCSSM.* (The quote is found on p. 84.)
89. Pang, J. (2013). Current elementary mathematics textbooks. In J. Kim, I. Han, M. Park, & J. Lee (Eds.), *Mathematics education in Korea: Curricular and teaching and learning practices* (pp. 43-61). Hackensack, NJ: World Scientific Publishing.
90. Pang, J. (2013). *Current elementary mathematics textbooks.*
91. Cuoco, A., Goldenberg, E., & Mark, J. (1996). Habits of mind: An organizing principle for mathematics curricula. *Journal of Mathematical Behavior, 15*, 375-402. Available at http://www.math.utah.edu/~ptrapa/math-library/cuoco/ Habits-of-Mind.pdf. (The quotes are found on pp. 376, 401.)
92. Partnership for Assessment of Readiness for College and Careers (PARCC). Available at http://www.parcconline.org.
93. The Smarter Balanced Assessment Consortium (Smarter Balanced). Available at http://www.smarterbalanced.org.
94. The National Center and State Collaborative Partnership. Available at http://www.ncscpartners.org.
95. The Dynamic Learning Maps Alternative Assessment System Consortium. Available at http://dynamiclearningmaps.org.
96. The University of Chicago School Mathematics Project (UCSMP). (n.d.). *Japanese and Russian textbook translations.* Available at http://ucsmp.u chicago.edu/resources/translations.

97. Singapore Math Inc. *Singapore textbook translations.* Available at http://www. singaporemath.com/default.asp.
98. gecko mathematics. Available at http://koreanmathematics.truman.edu.
99. Hoosiers Against Common Core. Available at http://hoosiersagainstcommonco re.com.
100. The cited reasons are found at Hoosiers Against Common Core. Retrieved from http://hoosiersagainstcommoncore.com/about-common-core-state-standa rds.
101. Usiskin, Z. (2011). *The common core: Where do we go from here?* (This presentation at the 2011 NCTM Annual Meeting can be retrieved from http://www.nctm.org/uploadedfiles/conferences/annual_meetings/webcasts/zusiski n%20nctm%202011%20ccs%20final.pdf.)
102. Fineout G., & Talley, T. (Mar. 2014). *Indiana withdrawing from common core standards.* Retrieved from http://bigstory.ap.org/article/indiana-withdraw ing-common-core-standards.
103. Korea Foundation for the Advancement of Science and Creativity. (2011). *Study of 2009 revision of the Korean mathematics curriculum.*

Printed in the United States
By Bookmasters